Particle emission concept and probabilistic consideration of the development of infections in systems

Marcus Hellwig

Particle emission concept and probabilistic consideration of the development of infections in systems

Dynamics from logarithm and exponent in the infection process, percolation effects

 Springer

Marcus Hellwig
Lautertal, Germany

ISBN 978-3-030-69499-9   ISBN 978-3-030-69500-2   (eBook)
https://doi.org/10.1007/978-3-030-69500-2

Responsible Editor: Reinhard Dapper
This Springer imprint is published by the registered company Springer Nature Switzerland AG
The registered company address is: Gewerbestrasse 11, 6330 Cham, Switzerland

Preface

Living beings are bound by the limited availability of energy on the planet on which they live. They need them on the one hand for their own life, on the other hand for the continuation of the species from which they come. The human being, as one of those, has penetrated with his demands on the availability of energy into areas from which he sees the continuation of the species as well as its growth ensured. The findings from the past show, however, that unrestricted growth reaches its limits when moderation—with consideration for the energy needs of other living beings—is not carried out. In this act of indefatigability, conflicts arise—on the one hand through the harassment of living beings of one's own species, on the other hand through the careless behavior when penetrating areas of life of other species. Other genera live in communities, the participants of which develop together in areas of life over a long period of time under defined population limits. This includes all participants in a symbiosis, including those who may have been transported outwards through whatever possibilities and coincidences. These places of residence of communities include, among other things, the organism of every mammal—including that of humans—including all organisms living in it, such as bacterial strains, which, for example, are indispensable in the act of digestion. In this community, there is agreement about task and effect, their advantages and disadvantages, and their balance should be undisturbed. Not so with the intrusion of something unknown, whose behavior can be unknown and sometimes harmful to the community. In this case, symbioses are equipped with recognition mechanisms that, if capable, detect strange particles and possibly render them harmless. Not all mechanisms are able to do this and therefore allow unrecognized persons to penetrate the area that should be protected, with the consequences that unknown periods of time can cause unknown damage without suffering damage themselves. This includes viruses, both from the biosphere and the digital sphere. The analysis of the past events of the virus pandemic will be part of the following work, it is reflected in the presentation of the particle emission concept and the application of prognoses about the frequencies of populations and their theoretical development considered via a probability function.

Cordial Thanks to

Mr. Edward Brown
United States Department of Health and Human Services
Department
Health Resources and Services Administration

sincerely, for his contributions, especially for:

- suggestions regarding potential implications for health care settings and epidemiological modeling, ratings, general suggestions, and for checking the English version.

This work was created with the excellent software Microsoft Office, the translations into the English language were mainly done by the Google translator with subsequent fine corrections.

What You Can Find In This Book

Claims

- it is not sufficient to only consider the spread of infection exponentially (particle emission concept),
- due to this fact the basis for the percolation is given, which excludes an infection of an individual with exactly one contact,
- because of the dynamics mentioned above, predictions can only be considered probabilistically,
- there is a consideration that based on the daily number of cases—at least suggests a number that refers to the group size that can be viewed as the source of an infection,
- It is possible to create a probabilistic preview that uses the mean 7-day logarithm of the case numbers in the past to determine an exponent for a probability density for a preview,
- From the considerations mentioned above it can be derived to what extent the forecasts for the
- Infection development and consequences for the strain on clinical capacities (personnel, equipment …) can endure.

Declaration of waiver

Due to the urgency of an early publication, the author waives his request to register and submit the thesis as a doctoral procedure.

Contents

Statement

1

The occurrence of events that have mutual influence on each other is the subject of this elaboration. People involved in these events are often amazed at the frequency with which they occur. One likes to speak of the coincidental in order to describe something like this as "fateful".

These types of events can be quantified when it comes to formulating a forecast.

These events include both "happy" and "unhappy" as well as any "coloration" between the two, if one wants to hide them under the guise of fate.

The effects that those involved have on one another are decisive for the outcome of the events.

In this book examples are listed which exert more or less "strong" mutual effects and what effects these have on the numerical development of the same.

This elaboration is a contribution to the development of the spread of epidemics; this includes the following basic assumptions and findings, which attribute every type of epidemic to a common systemic view.

Systemic Epidemics

2

From the author's point of view, a systemic distinction is made between epidemics of the following species:

- Technical epidemic—spread of a harmful virus in information systems,
- Biological epidemic—the spread of a harmful virus in living systems.

The aim is to recognize the effectiveness of an epidemic system and the possibility of foresight using statistical-probabilistic methods. For this it is necessary to clarify the systemic conditions.

The Occurrence of Events

3

3.1 Events (E)

The "occurrence of events" in this treatise is viewed as a period of time in which events take place together, albeit possibly offset in time within it.

What they all have in common is the fact that—at least 2 objects—participate in 1 event. "Objects" mean all those who can actually, actually, be involved in events that can cause causal effects.

For the purposes of our events, we will assume that each "object" is susceptible, that is to say the object is not yet infected and has the ability to be infected.

These include—to give examples—at least two system participants, (Fig. 3.1).

- System in motion
- System in contact
- System interaction

3.2 Risk and Opportunity (R, C)

But what makes that special about events?

The terms "risk and opportunity" come into play when a measure of the effect is to be defined.

In other words, the question is asked: "What happens if …?".

It is crucial that at least 2 participants are actually involved in 1 event E, Fig. 3.2, so that an effect can actually take place.

M. Hellwig, *Particle emission concept and probabilistic consideration of the development of infections in systems*,
https://doi.org/10.1007/978-3-030-69500-2_3

Fig. 3.1 System
participants

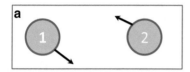 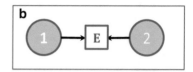

Fig. 3.2 A participants without -, 2.2.1 B—with participation in at least one event E

Then the question arises as to the "severity" of the effect, which is known as the risk.

Often only everyone, the risk R is described by E, but the chance C by E must also be rated equally.

By definition, both terms are accompanied by the probability of their occurrence P and the damage S or the gain G on a common, (Fig. 3.2), which are there:

$$R_E = S_E * P_E \tag{3.1}$$

$$C_E = C_E * P_E \tag{3.2}$$

If decisions are made, the risk and/or chance of a case is weighed up beforehand. This requires a comparison of the two values, from which a recommendation should emerge, so that the interactions in systems are considered.

Since the case under consideration involves risky events, the consideration of opportunities in the sense of advantageous events is excluded.

Fig. 3.2 Risk and chance
of an event for 2 participants

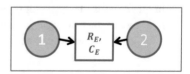

Interactions

4

4.1 Thought Sketch

The following formalism was developed from a simple thought sketch as it is present in Fig. 4.1. It arose from the simple question: "How often do glasses sound when the birthday is toasted with champagne?".

4.2 One-time Event that Occurs Within a Common Period of Time, the Infection, the Beginning of Percolations

If events and their opportunities and risks have previously been considered formally, what an event causes for those involved, biologically, in terms of information technology, will be described below (Fig. 4.2):

- the handshake causes the infection of a biological organism through skin moisture and its occupation with virus populations,
- Talking or singing causes the infection of a biological organism via respiratory gases and their penetration with virus populations.

There are no limits to the number of participants—without being hindered—they all have a formalism in common, which is listed as follows:

There are no limits to the number of participants—without being hindered—they all have a formalism in common, which is listed as follows:

1. In a set of participants, at least 2 participants (n = 2) share

© The Author(s), under exclusive license to Springer Nature Switzerland AG 2021   7
M. Hellwig, *Particle emission concept and probabilistic consideration of the development of infections in systems*,
https://doi.org/10.1007/978-3-030-69500-2_4

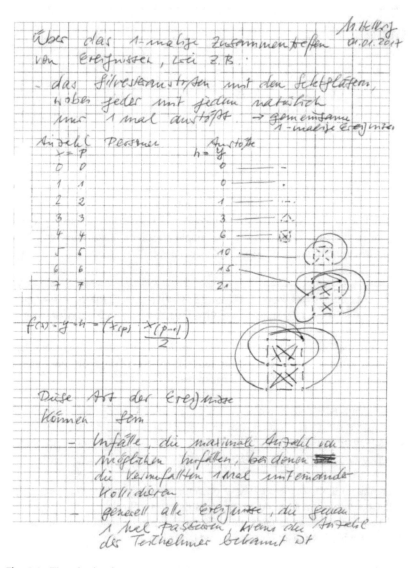

Fig. 4.1 Thought sketch

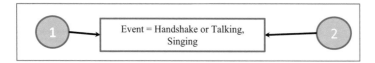

Fig. 4.2 Thought sketch

1 common event E,

it follows:

2. Are the events happening for everyone involved?

so is the number of events as a function:

$$A_E = (n_E * (n_E - 1))/2 \qquad (4.1)$$

or as a combination of the nth order, the case "without replacing", taking into account the arrangement:

$$A_E = n_E!/(2!(n_E - 2)!) \qquad (4.2)$$

4.2.1 Tabular Representation of the Development, the Distribution Rate

This results in the following tabular and graphical representation, Tables 4.1A and B with an increasing number of n participants who are involved in an event.

It can be seen from this that with a monotonously increasing number of participants n_E the number of events A_E increases square.

For illustration purposes, the first four cases (Fig. 4.3a, b, c, d) are shown with the help of figures.

It should then be determined that the limit value of

$$A_{E,2} = (n_E * n_E - 1)/2 \qquad (4.3)$$

$$(n_E * n_E - 1)/2 = \infty \qquad (4.4)$$

Table 4.1   A Number of frequent events, B

Number of participants	Combination number of common events	Number of common events function
Common, one-time events for which the following applies:	$A_E =$ $n_E!/(3!\,(n_E - 2)!/\,2)$	$A_E = (n_E * n_E - 1)/3$
0		0
1		0
2	0,5	0,5
3	1,5	1,5
4	3	3
5	5	5
6	7,5	7,5
7	10,5	10,5
8	14	14
9	18	18
10	22,5	22,5
11	27,5	27,5
12	33	33
13	39	39
14	45,5	45,5
15	52,5	52,5
16	60	60
17	68	68
18	76,5	76,5
19	85,5	85,5
20	95	95
21	105	105
22	115,5	115,5
23	126,5	126,5
24	138	138
25	150	150
26	162,5	162,5
27	175,5	175,5
28	189	189
29	203	203
30	217,5	217,5
31	232,5	232,5
32	248	248

A

common, double simultaneous events for which applies:

(chart showing the curve with plotted values: 0, 0, 0,5, 1,5, 3, 5, 7,5, 10,5, 14, 18, 22,5, 27,5, 33, 39, 45,5, 52,5, 60, 68, 76,5, 85,5, 95, 105, 115,5, 126,5, 138, 150, 162,5, 175,5, 189, 203, 217,5, 232,5, 248; x-axis 0 to 35, y-axis 0 to 300)

B

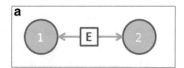

Fig. 4.3   a, b, c, d, number of common events in cases a, b, c, d. **a,** 2 participants, 1 event; **b,** 3 participants, 3 events; **c,** 4 participants, 6 events; **d,** 5 participants, 10 events

Fig. 4.3 (continued)

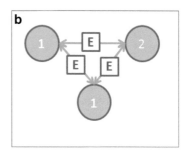

Fig. 4.3 (continued)

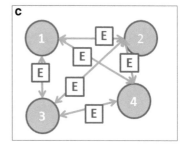

Fig. 4.3 (continued)

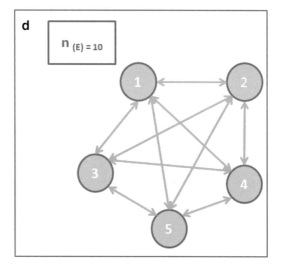

Then the limit for n = 5 is

$$(5 * (5 - 1))/2 = 10 \tag{4.5}$$

thus also the maximum of the possible highest number of events A_E is fixed. **This means that the risk of infection is 10 contacts over 5 participants = 200%**
Conversely, by changing from formula 3.2 to

$$n = \sqrt{(2 * A_{E,2} + 0{,}25) + 0{,}5)} \tag{4.6}$$

the number of infected people could be identified (biological, information technology) at the beginning of the survey.

4.3   Evolution of the Spread of Biological Infections, Distribution Rate and Contact Rate

Information technology infections caused by malicious viruses only cause material damage, which can be eliminated with appropriate systems. Biological damage can sometimes be permanent. The system under consideration therefore consists of the interaction of people under infectious conditions. Therefore, we fall back on the previously presented mathematical formalism, which is the propagation, hereinafter referred to as emission, under the functions:

$$A_{E,2} = (n_E * (n_E - 1))/2 \tag{4.7}$$

$$\text{now the distribution rate } V_{(e)} = (N_{(e)} * N_{(e-1)})/2 \tag{4.8}$$

$$\text{and the contact rate } K_{(e)} = N_{(e-1)}^{(r)} \tag{4.9}$$

4.3.1   Particulate Emission Concept

A concept, the particle emission concept, is presented that is intended to identify the functionality of the biological infection system.

To do this, it is first necessary to formulate the influencing variables, parameters of the system, which rare:

- The particles, in this case a number of harmful viruses,
- The emitter (s), the one who gives the particles,
- The number of issuers $N_{(e)}$
- The number of repetitions of contacts between at least 2 issuers (r),
- The contact rate $K_{(e)} = N_{(e-1)}^{(r)}$
- The rate of distribution $V_{(e)} = (N_{(e)} * N_{(e-1)})/2$

A hand sketch (Fig. 4.4) shows the concept of the functions used as follows:

Following the formalism, the following table (Fig. 4.5) results for issuers/participants with an increasing number and the risk of infection for recurrences, in such a way that they exchange infectious information with other participants.

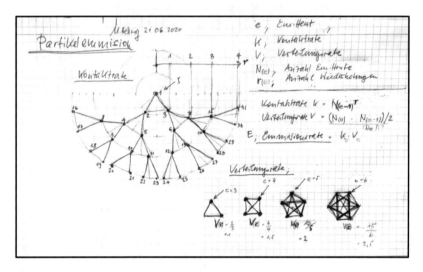

Fig. 4.4 Particle emission concept, hand sketch of the train of thought

4.3.2   Initial Conditions, Initial Population

At the beginning of an infection process, the way the infection behaves is often unknown, so it is necessary to make assumptions that—especially at the beginning of the infection process—must be confirmed or rejected by measurements for changes and thus corrected. The parameters (Fig. 4.6) can also change in the following process. This then repeatedly leads to corrections to the statements on the future of the course of the infection. The initial population can be shown in a table by rearranging the formula according to

$$n = \sqrt{((2 * V(e) + 0, 25)) + 0, 5} \qquad (4.10)$$

4.3.3   Finding the Initial Population

The determination of the initial population is part of an overall measurement concept; it must be carried out at the beginning of an infection process as well as in rhythmic, continuous time intervals across regions.

Development of the infection risk with an initial number of infected issuers (e), one-off possible contacts and repetitions (r)

Contact rate K(e) = (e-1)^r / Risk

Modell	Number of issuers involved (e)	Number of possible contacts — Distribution rate V(e)=1/2(e*(e-1))	Risk for r=0 (r=0)	Repeat case for r=1 (r=1)	Risk for r=1 (r=1)	Risk for r=2 (r=2)	Repeat case for r=2 (r=2)	Risk for r=3 (r=3)	Risk for r=3 (r=3)	Repeat case for r=4 (r=4)	Risk for r=4 (r=4)
	0,00										
	1,00										
	2,00	1,00	50%	1,00	100%	1,00	100%	1,00	100%	1,00	100%
	3,00	3,00	100%	6,00	200%	12,00	400%	24,00	800%	48,00	1600%
	4,00	6,00	150%	18,00	300%	54,00	900%	162,00	2700%	486,00	8100%
	5,00	10,00	200%	40,00	400%	160,00	1600%	640,00	6400%	2.560,00	25600%
	6,00	15,00	250%	75,00	500%	375,00	2500%	1.875,00	12500%	9.375,00	62500%
	7,00	21,00	300%	126,00	600%	756,00	3600%	4.536,00	21600%	27.216,00	129600%
	8,00	28,00	350%	196,00	700%	1.372,00	4900%	9.604,00	34300%	67.228,00	240100%
	9,00	36,00	400%	288,00	800%	2.304,00	6400%	18.432,00	51200%	147.456,00	409600%
	10,00	45,00	450%	405,00	900%	3.645,00	8100%	32.805,00	72900%	295.245,00	656100%
	11,00	55,00	500%	550,00	1000%	5.500,00	10000%	55.000,00	100000%	550.000,00	1000000%
	12,00	66,00	550%	726,00	1100%	7.986,00	12100%	87.846,00	133100%	966.306,00	1464100%
	13,00	78,00	600%	936,00	1200%	11.232,00	14400%	134.784,00	172800%	1.617.408,00	2073600%
	14,00	91,00	650%	1.183,00	1300%	15.379,00	16900%	199.927,00	219700%	2.599.051,00	2856100%
	15,00	105,00	700%	1.470,00	1400%	20.580,00	19600%	288.120,00	274400%	4.033.680,00	3841600%
	16,00	120,00	750%	1.800,00	1500%	27.000,00	22500%	405.000,00	337500%	6.075.000,00	5062500%
	17,00	136,00	800%	2.176,00	1600%	34.816,00	25600%	557.056,00	409600%	8.912.896,00	6553600%
	18,00	153,00	850%	2.601,00	1700%	44.217,00	28900%	751.689,00	491300%	12.778.713,00	8352100%
	19,00	171,00	900%	3.078,00	1800%	55.404,00	32400%	997.272,00	583200%	17.950.896,00	10497600%
	20,00	190,00	950%	3.610,00	1900%	68.590,00	36100%	1.303.210,00	685900%	24.760.990,00	13032100%

Fig. 4.5   a Particle emission concept, repeat cases of infectious exchange

Fig. 4.6 b Initial
population of infection with
an initial number of infected
people

Initial population of infection with an initial number of infected emitters (s)			
	Number of repetitions r as a function of time	Number of possible contacts	Initial population
Distribution rate $V_{(e)} = 1/2(e^*(e-1))$	$r_{(t)}$	$V_{(e=0)}$	e
	0,50	0,00	1
	0,50	0,00	1
	1,00	1,00	2
	1,50	3,00	3
	2,00	6,00	4
	2,50	10,00	5
	3,00	15,00	6
	3,50	21,00	7
	4,00	28,00	8
	4,50	36,00	9
	5,00	45,00	10
	5,50	55,00	11
	6,00	66,00	12
	6,50	78,00	13
	7,00	91,00	14
	7,50	105,00	15
	8,00	120,00	16

The measurements form the basis for statements on the future behavior of the infection process according to the applications of statistics and probabilities.

The figure below (Fig. 4.7) shows that for the case that at the beginning of a sample survey, the number of issuers (e) = 78 contributed to the distribution rate V (e) at least 13 emitters at a distance of r (e) = 6.5 repetitions. The number of repetitions can be shown as a time sequence using an example.

4.3.4   Determination of the Exponential Growth Over Subsequent Intervals

According to the presentation of the concept of particle emissions, however, it must also be considered that, depending on the density of the gatherings of the emitters, the case numbers assume exponential growth data.

A table (Fig. 4.8) shows how infections spread when the parameters assume the values listed.

Initial population of infection with an initial number of infected issuers (s)				
	Number of repetitions r as a function of time	Time units beginning e.g. 01/25/2020	Number of possible contacts	Initial population
Distribution rate $V_{(x)} = 1/2(e*(e-1))$	$r_{(t)}$	t	$V_{(s=0)}$	e
	0,50	25.1.20 0:00	0	1,00
	0,50	25.1.20 12:00	0	1,00
	1,00	26.1.20 0:00	1	2,00
	1,50	26.1.20 12:00	3	3,00
	2,00	27.1.20 0:00	6	4,00
	2,50	27.1.20 12:00	10	5,00
	3,00	28.1.20 0:00	15	6,00
	3,50	28.1.20 12:00	21	7,00
	4,00	29.1.20 0:00	28	8,00
	4,50	29.1.20 12:00	36	9,00
	5,00	30.1.20 0:00	45	10,00
	5,50	30.1.20 12:00	55	11,00
	6,00	31.1.20 0:00	66	12,00
	6,50	31.1.20 12:00	78	13,00
	7,00	1.2.20 0:00	91	14,00
	7,50	1.2.20 12:00	105	15,00
	8,00	2.2.20 0:00	120	16,00

Fig. 4.7 Example Chronological sequence of an infection with a conclusion about an initial population

For this purpose, 2 case study calculations according to the particle concept are listed:

Case study calculations according to the particle concept are listed:

Case 1: E (e) = 3, r = 4, K = 2^4, K = 3, V = 3; then E = K * V = 16*3 = 48,

Case 2: E (e) = 4, r = 4, K = 3^4, K = 3, V = 6; then E = K * V = 81*6 = 486.

The preceding explanation shows the manner in which the distribution of particles, i.e. infections via contacts and their contact rate, takes place. For a further prognosis of the further development, it is of decisive importance under which initial conditions, i.e. with which number of initially infected a development begins. Furthermore, it must be taken into account under which conditions the further course of an infection is learned. It should be noted that the spread dynamics would effectively change to something similar to N−1 if any of the group members were immune (for whatever reason). It gets more complicated, for example, when 4 out of 10 people are immune and 1 is infected and we cannot be sure that the 5 susceptible are all contacting the 1 infected or a person who is in contact with the infected and so on. But it is probably safe not to overwork this point. That means from this table is taken:

Development of the infection with an initial number of infected issuers (e), one-time possible contacts and repetitions (r)							Graphs
model	Number of issuers involved	Number of possible contacts	Repeat case for r =	Repeat case for r =	Repeat case for r =	Repeat case for r =	
			1	2	3	4	
	e	Distribution rate $V_{(e)} = 1/2(e*(e-1))$	r = 1	r = 2	r = 3	r = 4	
			Contact rate $K(e) = (e-1)^r$				
	0,00	0,00	-6,00	6,00	-6,00	6,00	
	1,00	0,00	0,00	0,00	0,00	0,00	
	2,00	1,00	1,00	1,00	1,00	1,00	
	3,00	3,00	6,00	12,00	24,00	48,00	
	4,00	6,00	18,00	54,00	162,00	486,00	
	5,00	10,00	40,00	160,00	640,00	2.560,00	
	6,00	15,00	75,00	375,00	1.875,00	9.375,00	
	7,00	21,00	126,00	756,00	4.536,00	27.216,00	
	8,00	28,00	196,00	1.372,00	9.604,00	67.228,00	
	9,00	36,00	288,00	2.304,00	18.432,00	147.456,00	
	10,00	45,00	405,00	3.645,00	32.805,00	295.245,00	
	11,00	55,00	550,00	5.500,00	55.000,00	550.000,00	
	12,00	66,00	726,00	7.986,00	87.846,00	966.306,00	
	13,00	78,00	936,00	11.232,00	134.784,00	1.617.408,00	
	14,00	91,00	1.183,00	15.379,00	199.927,00	2.599.051,00	
	15,00	105,00	1.470,00	20.580,00	288.120,00	4.033.680,00	
	16,00	120,00	1.800,00	27.000,00	405.000,00	6.075.000,00	
	17,00	136,00	2.176,00	34.816,00	557.056,00	8.912.896,00	
	18,00	153,00	2.601,00	44.217,00	751.689,00	12.778.713,00	
	19,00	171,00	3.078,00	55.404,00	997.272,00	17.950.896,00	
	20,00	190,00	3.610,00	68.590,00	1.303.210,00	24.760.990,00	

Fig. 4.8 Table for the particulate emission concept, case 1, case 2

1. The number of infected people involved, which is taken into account at the time the recording begins,
2. The number of infectious contacts that arise from 1. that are to be expected from the formula listed at the time of the first contact—at the beginning of the recording,
3. The number of infectious contacts resulting from the recurrence cases applies if the infections can spread according to the corresponding exponent for the recurrence/replication (r).

4.4 Basis for a Probabilistic Prognosis

A statement about the progress of the infection process that develops after the initial condition has been determined can be presented using probability profiles.

For this purpose, a skewed distribution is used in this book, as it is assumed that a long-term frequency distribution is expected due to the saturation of the infection process along a time axis. A probabilistic statement should help to show the maximum as well as the end of the infection process on a time axis.

4.4.1 Statistical Surveys

Statistical surveys, such as those given using the example of the corona infection in 2020 on the corresponding website https://de.statista.com, serve as the basis for the frequency distributions.

Using the example of the infection events in Germany, the United States of America and Spain, it is shown how the initial development is.

Statistical surveys Germany (Fig. 4.9).

4.4.2 Probability

To determine the future occurrence of events that are statistically recorded, the probability can make statements. For this purpose, measurements according to Sect. 3.4.1 are required, which on the one hand provide the basis for the starting conditions of the probability and for a continuous forecast.

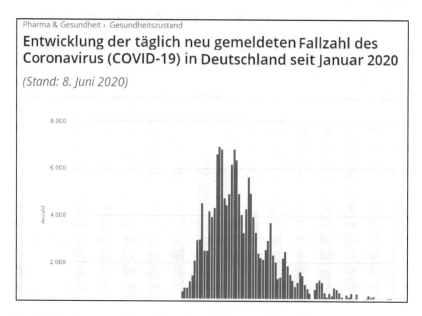

Pharma & Gesundheit › Gesundheitszustand

Entwicklung der täglich neu gemeldeten Fallzahl des Coronavirus (COVID-19) in Deutschland seit Januar 2020

(Stand: 8. Juni 2020)

Fig. 4.9 Development of the number of cases of the coronavirus (COVID-19) reported daily in Germany since January 2020

4.4.3   The Difference: Mathematical Truth Through Proof—Statistical Approximation to Truth Through Experiments

If scientists try to find the truth—in the sense of—100 percent certainty—they will fail. There will be a difference between mathematical and statistical truth at all times.

A mathematical truth is defined as a mathematical proof.

A statistical truth, for which a proof can never be found in a mathematical point of view, is only valid as a comparison of samples of a series of experimental values with a theoretical density function, which always depends on the amount of experiments they collect.

Ultimately, the answer is: On the one hand, the strength of a deterministic algorithm in the mathematical sense is used, and on the other hand, a quantity of data is evaluated in the statistical sense using a density distribution such as the logarithmic density distribution to approximate a truth.

Fig. 4.10 Proof of the
Pythagorean theorem

Pythagorean theorem
Numerical example as proof

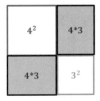

$49\ cm^2 - 24\ cm^2 = 25\ cm^2$

$49\ cm^2 - 24\ cm^2 = 25\ cm^2$

$\sqrt{25cm^2} = 5\ cm$

In connection with the preceding explanations, it is pointed out that this work only makes statistical-probabilistic statements, influences from other specialist areas are not taken into account.

An example of the elaboration of a mathematical truth is the proof of the Pythagorean theorem. The one that is often published is this graphic representation of a numerical proof (Fig. 4.10) in which the truth is brought about by a compelling logic case.

In contrast, there is the statistical-probabilistic finding of truth, the approximation to a correspondence of relationships by means of a regression test, which is brought about by the method of least squares. This method is used in the further course of the comparison of the frequency values of infection values and the probability values Fig. 4.11a, b) from the logarithmic equibalanced distribution. The percentage of the determined coefficient of determination is therefore to be regarded as an approximate value for agreement.

Since it was expected and shown that the statistical frequency distribution of the infection process is not symmetrical, the density shown below was used for forecasts:

4.5   Doubts About Statistical Measurements

Many questions revolve around the value of the statistical surveys, so one of the important ones is:

- Is the number of coronavirus (SARS-CoV-2) cases increasing because more people are being tested?

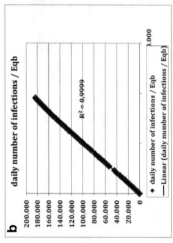

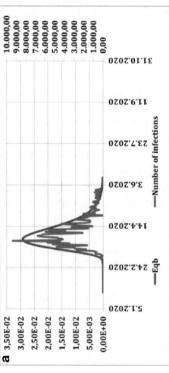

Fig. 4.11 **a** Frequency values—probability values, **b** method of least squares

This can definitely be answered in the sense that all test measurements that were collected per quarter are related to the numbers of people who tested positive. The following table (Fig. 4.12) shows that the deviations from quarter to quarter are in the lower percentage range. This secures the basis for all statistical surveys.

All statistical surveys are subject to fixed rules so that the key figures determined from them always remain traceable and so that repetitions can always be carried out under consistently high standards.

	Week	2020	Number		
	Testings				
	positively	tested	prositive-ratio	(%)	Number of laboratoies
	Bis	einschließlich	KW10	124.716	3.892
					Deviation from the previous measurement
	Measurement	Mean	Mean	Coeffizient	measurement
1	Woche 11-14	311.484,8	24.925,3	8,00%	
2	Woche 15-18	374.669,8	30.727,5	8,20%	0,1991%
3	Woche 19-22	370.490,5	30.293,0	8,18%	-0,0248%
4	Woche 23-26	371.084,3	26.960,3	7,27%	-0,9112%
5	Woche 27-30	350.694,3	20.891,0	5,96%	-1,3082%
1	11	127.457	7.582	5,9	114
2	12	348.619	23.820	6,8	152
3	13	361.515	31.414	8,7	151
4	14	408.348	36.885	9	154
1	15	380.197	30.791	8,1	164
2	16	331.902	22.082	6,7	168
3	17	363.890	18.083	5	178
4	18	326.788	12.608	3,9	175
1	19	403.875	10.755	2,7	182
2	20	432.666	7.233	1,7	183
3	21	353.467	5.218	1,5	179
4	22	405.269	4.310	1,1	178
1	23	340.986	3.208	0,9	176
2	24	326.645	2.816	0,9	172
3	25	387.484	5.309	1,4	175
4	26	467.004	3.674	0,8	180
1	27	505.518	3.080	0,6	150
2	28	509.298	2.989	0,6	177
3	29	537.334	3.480	0,6	173
4	30	569.868	4.462	0,8	176
	31	573.802	5.551	1	161
	Summe	8.586.648	249.242		

Fig. 4.12 Frequency values—probability values, **b** method of least squares

The Difference Between Influenza and COVID Waves

Graphics showing the waveforms of infections are presented to the public. The forms result from the statistical surveys, which show the number of infections registered over a period of time—it is frequency distributions that assume waveforms. It should be noted that there are fundamental differences between an influenza wave and a COVID wave.

These are the most obvious:

- An influenza wave of flu develops on the left flat and moderately—steep right (Fig. 5.1)

- A COVID wave develops on the left and flat on the right (Fig. 5.2).

Further differences/similarities that outline the nature of the infections can be seen when the recurrences are observed over a number of years, these are:

1. Influenza flu waves
   a) go hand in hand with the cold seasons
   b) have a time-limited area
   c) have a determinable period of time until the clinical picture appears
   d) are—due to the controllable number of cases—clearly registrable and traceable
   e) can be influenced by vaccinations
2. COVID waves
   a) are independent of the seasons
   b) do not have a time-limited area, but can be repeated
   c) have no ascertainable period of time until the clinical picture appears

M. Hellwig, *Particle emission concept and probabilistic consideration of the development of infections in systems*,
https://doi.org/10.1007/978-3-030-69500-2_5

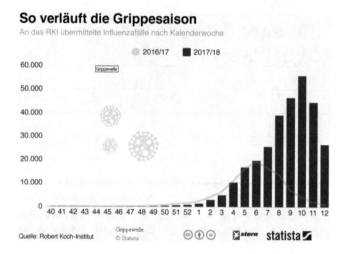

Fig. 5.1   Influenza cases 2017/2018

(Stand: 2. November 2020)

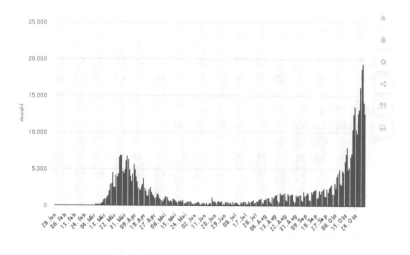

Fig. 5.2   COVID case numbers 2020

d) are—if the number of cases is exceeded—no longer registerable and no longer traceable

e) can be influenced by vaccinations

The consequences with which a COVID infection process is more risky than the influenza flu wave can be derived from this.

- Due to the influenza flu waves that occur at regular intervals every season, the risk of a COVID wave is seen in the process of getting used to one. It should subside after a certain time, it can be influenced by drugs, further infections no longer occur.
- The fallacy is expressed in the assumption that the COVID wave behaves in the same way.
- This leads to another wave starting up if it is not ensured that a general block to spread is set up early on.
- It also leads to a further fallacy that—after the infection has occurred—the steepness of a wave curve can be converted into a flatness "flatten the curve". This is not the case, because the initial steepness is maintained until the incline between at least two points on the incline leads to a change in the subsequent course. Rather, the curve runs according to a tie or an increase or decrease in the number of cases. In this respect, the course of the COVID curve is left-hand and right-hand flat (Fig. 5.3).

The aforementioned properties lead to subjecting the COVID infection as such to a differentiated statistical-probabilistic view.

The New York Times

One chart explains why slowing the spread of the infection is
nearly as important as stopping it.

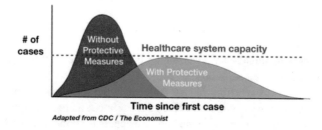

Fig. 5.3 Flattening the coronavirus curve

Limits of Symmetrical Variance

6

The manifestations of frequency distributions, as they are evident in almost all subject areas, influence the objective recording of situations in that they are often used as a basis for judgment. The process world also likes to use simple, memorable graphic representations. The symmetrical normal distribution density developed by Gauss is a good example of this. On the other hand, there are numerous asymmetrical process positions for which specially adapted density functions were developed.

The equibalance distribution Eqbl, which has been expanded to objectify the slope / kurtosis, is intended to remedy the situation by replacing as many of the specially adapted density functions as possible via a skew parameter as well as a logarithmic influence—the fourth parameter.

For the quality-effective monitoring and action management, the newly developed formula of a right or left skewed distribution, combined with the kurtosis, the "Equibalancedistribution Eqbl" for the analysis of measured values is presented as a theoretical variant Eqbl is no longer included as a simplified special case, but adjusts itself with its logarithmic portion to the conditions of multiplicative influences from the raw data.

The same applies to the Eqbl: It is the case, however, that due to the mutual influence of the parameters on the values that the Eqbl delivers, it will not be possible to estimate individual parameters with conventional statistics because they all already occur in the expected value.

First, however, the analysis of the basic form of the logarithmic Equibalancedistribution presented later is presented: the Equibalancedistribution (Eqb).

M. Hellwig, *Particle emission concept and probabilistic consideration of the development of infections in systems*,
https://doi.org/10.1007/978-3-030-69500-2_6

6.1   Analysis of the Eqb Density

Symmetrical forms of appearance, as they are revealed in almost all specialist areas, influence the objective recording of facts in such a way that they are often used as a basis for judgment. The process world also likes to use simple, memorable graphical representations.

The symmetrical normal distribution density developed by Gauss is a good example of this. On the other hand, there are numerous asymmetrical process positions for which specially adapted density functions were developed.

The newly developed equibalance distribution Eqb is intended to remedy the situation by replacing as many of the specially adapted density functions as possible via a skew parameter.

The newly developed formula of a right or left skewed distribution, the "Equibalancedistribution Eqb" for the analysis of measured values, is trend-setting for the quality-effective monitoring and action management.

The symmetrical normal distribution used to describe the description is still contained in the Eqb as a simplified special case.

It is, however, the case that due to the mutual influence of the parameters on the values that the Eqb delivers, it will not be possible to estimate individual parameters with conventional statistics, because they all already occur in the expected value.

The mathematical function Equibalancedistribution Eqb is examined (German language):

$$\frac{1}{\sqrt{2\pi\sigma^2(1 - \rho(x - \mu))}} \exp -\frac{(x - \mu)^2}{2\sigma^2(1 - \rho(x - \mu))} \tag{6.1}$$

Eine Familie von Verteilungen auf $\mathbb{R}$

Andrej Depperschmidt und Marcus Hellwig

1. August 2016

Zusammenfassung
Wir betrachten eine parametrische Familie von Funktionen auf $\mathbb{R}$, die die Dichten der Normalverteilungen enthalten. Wir zeigen, dass alle Funktionen in dieser Familie selbst Dichten von Verteilungen sind.

1 Familie von Dichten

Für $r \in \mathbb{R}$, $\mu \in \mathbb{R}$ und $\sigma^2 > 0$ betrachten wir die Funktionem $f_{\rho;\mu,\sigma^2} : \mathbb{R} \to \mathbb{R}$ definiert durch

$$f_{\rho;\mu,\sigma^2}(x) = \begin{cases} \frac{1}{\sqrt{2\pi\sigma^2(1-\rho(x-\mu))}} \exp\left\{-\frac{(x-\mu)^2}{2\sigma^2(1-\rho(x-\mu))}\right\} & : x < 1/\rho + \mu \\ 0 & : x \geq 1/\rho + \mu. \end{cases}$$

In dem Fall $\rho = 0$ stimmt $f_{\rho;\mu,\sigma^2}(x)$ mit der Dichte der Normalverteilung mit Parametern μ und σ^2 überein. (Wir fassen zur Konsistenz 1 / 0 als ∞ auf.

Wir beschränken uns in der Analyse der Funktion zunächst auf den Fall $\mu = 0$ und $\sigma^2 = 1$ und setzen $f_\rho = f_{\rho;0,1}$.

Theorem 1 *Die Familie $\{f_\rho : \rho \in \mathbb{R}\}$ist eine Familie von Dichten von Wahrscheinlichkeitsverteilungen auf $\mathbb{R}$.*

(i)   *Für $\rho = 0$handelt es sich bei der Verteilung um die Standardnormalverteilung.*

(ii) *(Für $\rho > 0$ist die Verteilungsfunktion gegeben durch*

$$F_\rho^+(x) = \begin{cases} \Phi\left(\frac{x}{\sqrt{1-\rho x}}\right) + e^{2/\rho^2}\Phi\left(\frac{x}{\sqrt{1-\rho x}} - \frac{2}{\rho\sqrt{1-\rho x}}\right) & : x < 1/\rho, \\ 1 & : x \geq 1/\rho. \end{cases} \qquad (6.1.1)$$

(iii) *Für $\rho < 0$ist die Verteilungsfunktion gegeben durch*

$$F_\rho^-(x) = \begin{cases} 0 & : x \leq 1/\rho, \\ \Phi\left(\frac{x}{\sqrt{1-\rho x}}\right) + e^{2/\rho^2}\Phi\left(\frac{x}{\sqrt{1-\rho x}} - \frac{2}{\rho\sqrt{1-\rho x}}\right) & : x > 1/\rho. \end{cases} \qquad (6.1.2)$$

Beweis. Für jedes $\rho \in \mathbb{R}$ ist f_ρ nicht negativ. Für $\rho > 0$ ist f_ρ auf dem Intervall $(-\infty, 1/\rho)$ definiert. Für $\rho < 0$ ist die Funktion auf dem Intervall $(1/\rho, +\infty)$ definiert. Für $\rho = 0$ handelt es sich bei f_ρ um die Dichte der Standardnormalverteilung.

Wir zeigen nun, dass f_ρ für jedes $\rho \in \mathbb{R}$ die Dichte einer Wahrscheinlichkeitsverteilung auf $\mathbb{R}$ ist. Wir bezeichnen im Folgenden mit φ und Φ die Dichte beziehungsweise die Verteilungsfunktion der Standardnormalverteilung. Damit ist (i) klar.

Für (ii) und (iii) ist zu zeigen, dass f_ρ jeweils die Ableitung F_ρ^+ und von F_ρ^- ist und dass beide letztere Funktionen Verteilungsfunktionen sind. Offensichtlich sind die Funktionen stetig und monoton wachsend auf $\mathbb{R}\backslash\{1/\rho\}$.

(ii) Betrachten wir zunächst den Fall $\rho > 0$. Es gilt

$$\lim_{x \nearrow 1/\rho} F_\rho^+(x) = \lim_{x \nearrow 1/\rho} \left(\Phi\left(\frac{x}{\sqrt{1-\rho x}} \right) + e^{2/\rho^2} \Phi\left(\frac{x}{\sqrt{1-\rho x}} - \frac{2}{\rho\sqrt{1-\rho x}} \right) \right)$$

$$= \Phi(\infty) + e^{2/\rho^2} \Phi(-\infty) = 1 + 0.$$

Also ist F_ρ^+ stetig in $1/\rho$ und damit auf ganz $\mathbb{R}$. Ferner gilt $\lim_{x\to-\infty} F_\rho^+(x) = 0$.

Durch Ableiten nach x überzeugt man sich leicht davon, dass F_ρ^+ die Verteilungsfunktion einer Wahrscheinlichkeitsverteilung auf $\mathbb{R}$ ist, deren Dichte durch f_ρ gegeben ist. Es gilt nämlich

$$\frac{d}{dx} F_\rho^+(x) = \left(\frac{\rho x}{2(1-\rho x)^{3/2}} + \frac{1}{(1-\rho x)^{1/2}} \right) \varphi\left(\frac{x}{\sqrt{1-\rho x}} \right)$$

$$- e^{2/r^2} \frac{\rho x}{2(1-\rho x)^{3/2}} \varphi\left(\frac{x}{\sqrt{1-\rho x}} - \frac{2}{\rho\sqrt{1-\rho x}} \right)$$

$$= f_\rho(x) + \frac{\rho x}{2(1-\rho x)^{3/2}} \left(\varphi\left(\frac{x}{\sqrt{1-\rho x}} \right) - e^{2/\rho^2} \varphi\left(\frac{x}{\sqrt{1-\rho x}} - \frac{2}{\rho\sqrt{1-\rho x}} \right) \right)$$

$$= f_\rho(x).$$

Hier haben wir benutzt, dass $f_\rho(x) = \frac{1}{\sqrt{1-\rho x}} \varphi\left(\frac{x}{\sqrt{1-\rho x}} \right)$ ist. Die letzte Gleichung im Display folgt wegen

$$\varphi\left(\frac{x}{\sqrt{1-\rho x}} \right) - e^{2/\rho^2} \varphi\left(\frac{x}{\sqrt{1-\rho x}} - \frac{2}{\rho\sqrt{1-\rho x}} \right)$$

$$= \frac{1}{\sqrt{2\pi}} \left(\exp\left\{ -\frac{x^2}{2(1-\rho x)} \right\} - \exp\left\{ \frac{2}{\rho^2} - \frac{1}{2}\left(\frac{x^2}{1-\rho x} - \frac{4x}{\rho(1-\rho x)} + \frac{4}{\rho^2(1-\rho x)} \right) \right\} \right)$$

$$= \frac{1}{\sqrt{2\pi}} \exp\left\{ -\frac{x^2}{2(1-\rho x)} \right\} \left(1 - \exp\left\{ \frac{2}{\rho^2} + \frac{2x}{\rho(1-\rho x)} - \frac{2}{\rho^2(1-\rho x)} \right\} \right)$$

$$= \frac{1}{\sqrt{2\pi}} \exp\left\{ -\frac{x^2}{2(1-\rho x)} \right\} (1 - e^0) = 0.$$

(iii) Betrachten wir nun den Fall $\rho < 0$. Es gilt

$$\lim_{x \searrow 1/\rho} F_\rho^-(x) = \lim_{x \searrow 1/\rho} \left(\Phi\left(\frac{x}{\sqrt{1-\rho x}} \right) + e^{2/\rho^2} \Phi\left(\frac{x}{\sqrt{1-\rho x}} - \frac{2}{\rho\sqrt{1-\rho x}} \right) \right)$$

$$= \Phi(-\infty) + e^{2/\rho^2} \Phi(-\infty) = 0.$$

Also ist F_ρ^- stetig in $1/\rho$ und damit auf ganz $\mathbb{R}$. Dass f_ρ die Ableitung von F_ρ^- ist, zeigt man analog zu dem Fall $\rho > 0$.

Lemma 1.1 *Für jede Nullfolge* (ρ_n) *gilt*

$$e^{1/\rho_n^2}\,\Phi(-1/|\rho_n|^{3/2}) \overset{n\to\infty}{\longrightarrow} 0. \tag{6.1.3}$$

Beweis. Wir verwenden die folgende Abschätzung

$$1 - \Phi(x) \le \frac{\phi(x)}{x} \ \text{für alle } x > 0.$$

Damit erhalten wir

$$e^{1/\rho_n^2}\,\Phi\big(-1/|\rho_n|^{3/2}\big) = e^{1/\rho_n^2}\big(1 - \Phi\big(1/|\rho_n|^{3/2}\big)\big) \le |\rho_n|^{3/2}\,\frac{1}{\sqrt{2\pi}}\,\exp\bigg\{\frac{1}{\rho_n^2} - \frac{1}{2|\rho_n|^3}\bigg\} \overset{n\to\infty}{\longrightarrow} 0.$$

Lemma 1.2 *Für* $\rho \to 0$ *konvergieren* F_ρ^+ *und* F_ρ^- *schwach gegen* Φ. *Mit anderen Worten gilt*

$$\lim_{\rho \searrow 0} F_\rho^+(x) = \lim_{\rho \nearrow 0} F_\rho^-(x) = \Phi(x) \quad \text{für alle } x \in \mathbb{R}.$$

Beweis. Sei $x \in \mathbb{R}$ und sei (ρ_n) eine Folge mit $\rho_n > 0$ für alle n und $\rho_n \to 0$ für $n \to \infty$. Für genügend große n ist dann $x < 1/\rho_n$ und es gilt

$$F_{\rho_n}^+(x) = \Phi\bigg(\frac{x}{\sqrt{1 - \rho_n x}}\bigg) + e^{2/\rho_n^2}\,\Phi\bigg(\frac{x}{\sqrt{1 - \rho_n x}} - \frac{2}{\rho_n\sqrt{1 - \rho_n x}}\bigg).$$

Der erste Summand konvergiert für $n \to \infty$ gegen $\Phi(x)$. Der zweite Summand verschwindet nach Lemma 1.2.

Analog sieht man, dass für jedes $x \in \mathbb{R}$ und für jede Nullfolge (ρ_n) mit $\rho_n < 0$ für alle n die Folge $F_{\rho_n}^-(x)$ gegen $\Phi(x)$ konvergiert.

6.2   Adding the Kurtosis Parameter to the Density Eqb4

It became evident that statistical surveys and the resulting frequency distributions are often not symmetrical with regard to the scatter around an expected value or mean value. Rather, the values tend to a maximum with the consequence of a manifestation of skewness and kurtosis.

The following equation, density, was used in the following evaluations.

$$Eqb4(x; \delta, md, r, \kappa) = \left(\frac{1}{s * \sqrt{\left(2\pi\left(\frac{1-((r)*(x-md))}{\kappa}\right)\right)}} * EXP\left(\left(-\left(\frac{1}{2} * \frac{\left(\frac{x-md}{s}\right)^2}{1-(r*(x-md))}\right) * \kappa\right)\right) \right.$$

(6.2)

6.2.1   Parameter Estimation

As a result, the parameter estimates for

- mean estimate = modal
- and estimation of the spread:

$$\hat{\sigma}^2 = s_n^2 = \frac{1}{n-1} \sum_{i=1}^{n} (x_i - \overline{x})$$

(6.3)

- as well as the estimated skewness of the sample values according to:

$$\hat{v} = \frac{1}{n} \sum_{i=0}^{n} ((x_i - \overline{x})/s)^3$$

(6.4)

- as well as the estimated kurtosis of the sample values

$$\hat{k}_{,} = \left(\frac{n(n+1)}{(n-1)(n-2)(n-3)} \sum_{i=0}^{n} ((x_i - \overline{x})/s)^4 \right) - \frac{3(n-1)^2}{(n-2)(n-3)}$$

(6.5)

The values of the skewness of the modeled Eqb and the skewness of the sample should—since this is an approximation method—differ by small differences—uncertainty.

Fig. 6.1 Temporal
sequence, date and
cumulative number of
infected people

30.4.2020	162.123
29.4.2020	160.059
28.4.2020	159.038
27.4.2020	159.142
26.4.2020	157.026
25.4.2020	155.418
24.4.2020	153.215
23.4.2020	151.175
22.4.2020	148.453
21.4.2020	147.065
20.4.2020	145.743
19.4.2020	144.348
18.4.2020	142.465

The aforementioned parameters must be obtained from the measurement data of the respective measurement system. The present application is about measurement data obtained from the statista.com website. An excerpt from a temporal sequence serves as an example:

This list (Fig. 6.1) deals with measurement data obtained from the website statista.com. An excerpt from a chronological sequence serves as an example:

6.3   Forecast using the density function and continuous adjustment of the parameters

The system under consideration is a dynamic system that shows changes in the effects on the future of the system in accordance with the change in the parameters from the measurement data.

6.3.1   Statistical Basis

For this purpose, the aforementioned parameters are described in their effect on a course to be predicted with regard to the application of the equibalance distribution.

Fig. 6.2 Skewed
distribution

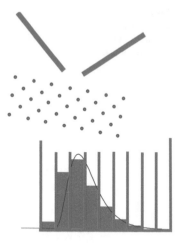

- The mode value: it shows where (in terms of time or number) the maximum can be expected,
- The scatter value, the standard deviation: shows how far the measurement data scatter around the mode value,
- The skewness: it shows to what extent the ends of the measurement data are distributed over the spread "to the left or to the right" around the mode value.

A normal distribution cannot provide complete values for this because it does not take the skew parameter into account, so a statement about the end of an infection process cannot be made (6.2).

6.3.2   Basics for the Exponential Expansion

All parameters for a density are determined from the data of the determined population or a sufficiently large number of data from a sample. Infection processes with an exponential spread are often described.

But what is the exponent in a dynamic infection process?

The following determinations are based on the idea of connecting the exponential development with the history of the logarithmic values—in other words:

- The exponent values of a limited development in the future are determined from a limited number of logarithmic values from history.

This makes it possible to create a prognosis that can be used to change the behavior of the causer / participant in the infection process.

An example based on test data (Fig. 6.3, 6.4, 6.5) from the USA may prove this.

At one point in time, the test data are listed as a frequency distribution as well as the values of the probability function. It can be seen that under the existing exponential development, a decrease in accordance with the probability function is to be expected.

At a later point in time, the test data are listed as a frequency distribution as well as the values of the probability function. It can be seen that under the existing exponential development, a decrease in accordance with the probability function is to be expected.

The **theoretical prognosis from the determination of the historical logarithm** indicates a change in the exponential future course.

If the graphic is compared with the **theoretical forecast from the determination of the historical logarithm** and the **existing exponential development**, then there are certain indications that an increase in frequency is to be expected—even if not in complete agreement.

As described, all parameters mentioned are subject to the dynamics of the system under consideration.

For this purpose, it is measured how the measurement data, in this case the:

- Differentiate the daily number of cases from day to day.

Thus, the values to be determined fall into the use of the exponential function of the type:

$$\text{general: } f_{(x)} = b^x, b \in \mathbb{R}\backslash\{1\} \tag{6.3.1}$$

and its inverse function for determining the exponent according to.

$$\text{general: } b^x = a; \quad x = log_b\, a \tag{6.3.2}$$

his results from the average daily mean logarithm of the previous 7 days:

$$f_{(x)} = b^x,\, for\; x = LOG(mean(x/days);\, mean(measurements))$$

The category represents the interval that is necessary for the comparison of frequency distribution and density function.

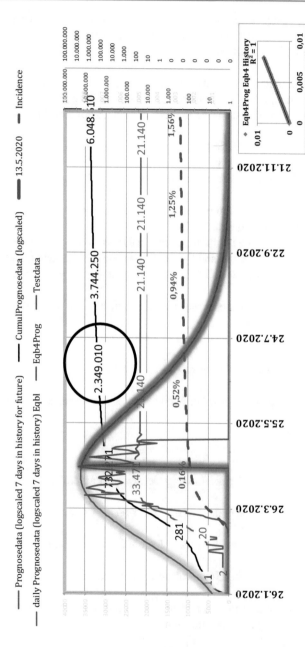

Fig. 6.3 Frequency and probability density without prognosis from the log. development

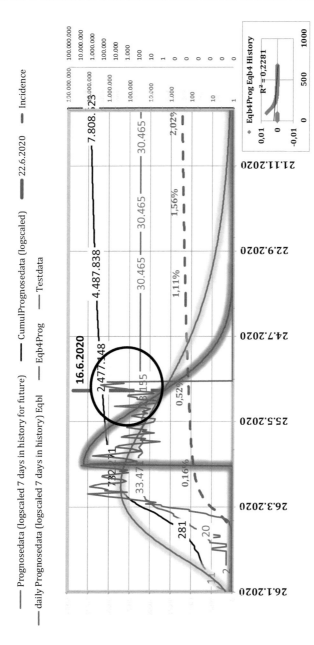

Fig. 6.4 Frequency and probability density with a 21-day prognosis from the log. development

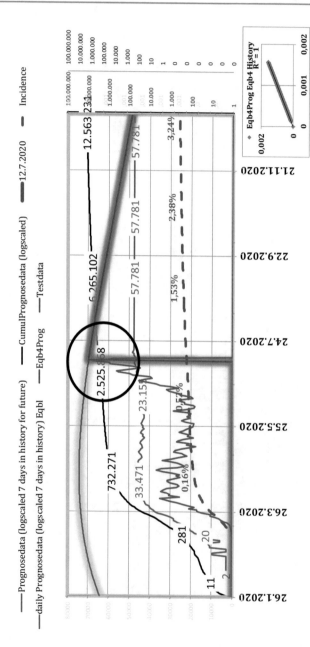

Fig. 6.5   Existing exponential development

6.4   Data Analysis on the Particle Emissions Concept

The preceding examples should show:

1. An infection process is a dynamic process that is subject to developments that can be influenced
2. The evaluation of an infection process depends on the quality of the measurements with regard to:
   (a) the sequence of measurements, timing,
   (b) the number of measurements within the sequence.
3. The evaluation of the measurements and a prognosis for the future of an infection is subject to the particle emission concept:
   (a) The particles, in this case a number of harmful viruses, are subject to the particle emission concept,
   (b) The emitter (s), the one who gives the particles,
   (c) The number of issuers $N_{(e)}$.
   (d) The number of repetitions of contacts between at least 2 issuers (r),
   (e) The contact rate $K_{(e)} = N_{(e-1)}^{(r)}$.
   (f) The rate of distribution $V_{(e)} = (N_{(e)} * N_{(e-1)})/2$.

6.4.1   Consequences of Hygiene, Handshake, Breathing Air (Aerosols)

Since infections can spread more than exponentially under the observed and measured conditions, as shown, the noticeable beginnings are to be controlled immediately by continuous measurements and subsequently to be dampened by suitable methods. Examples of contact and distribution, beginning of an infection "handshake" (Fig. 6.6).

The increasing behavior of the process, which was not recognized at an early stage, can then only be influenced with difficulty, since—as shown—the spread develops more than exponentially over time.

This applies to all types of events that lead to the emission of viruses: transmission from fluids such as body fluids from handshakes or particles in the breath.

According to the particle emission concept, the risk of infection increases exponentially several times for a group of people in a common location (Fig. 6.7).

The statistics and probability analysis, from which it is possible to derive what the forecast of the distribution over time will look like, is dependent on the values

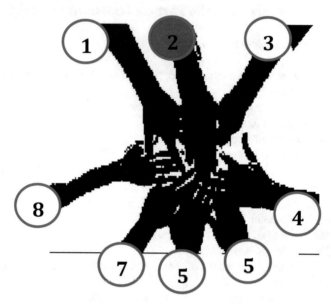

Fig. 6.6 Infection 1 issuer, 7 participants

from the particle emission concept, as the categorization of the current and future distributions is derived from them.

It is obvious that virus particles spread many times more intensely when they are exhaled than when they come into contact with one another, such as a handshake.

Therefore, the exponential rate of increase via aerosols also increases many times over, depending on the number of participants in the event and the duration of the exposure, primarily where people like—close—to get together:

- wedding parties
- Singing and trombone clubs
- department stores
- Bars and pubs
- ...

The reader may check for himself where he has been in close association with other people in the past.

Fig. 6.7 Emitters of the infection 3 participants

6.4.2   Determination of the Prognosis for a Future Increase or Decrease in the Infection Rate

The infection process is a dynamic process that can be influenced by:

- The change in the values for the parameters K and V.

Any changes in the values in the course of the current process, be they favorable or unfavorable, change the aforementioned logarithm and thus influence the further exponentially calculated further course of the future of the event (Fig. 6.8 until 6.14).

The statement on this is supported by the following diagrams for the course of the infection in Germany with the statuses: 03/18/2020, 03/25/2020, 04/02/2020, 04/10/2020, 05/01/2020, 05/14/2020, 05/30/2020.

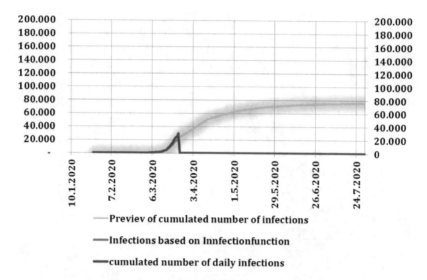

—Previev of cumulated number of infections

—Infections based on Innfectionfunction

—cumulated number of daily infections

Fig. 6.8 Infection events as of March 18, 2020, preview of future total development, infections based on function, total development of daily infections

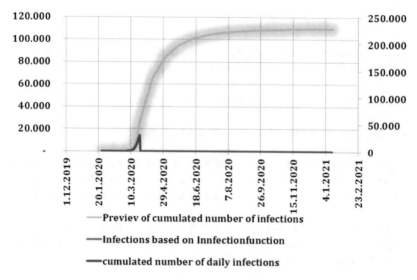

—Previev of cumulated number of infections

—Infections based on Innfectionfunction

—cumulated number of daily infections

Fig. 6.9 Infection events as of March 25, 2020, preview of future total development, infections based on function, total development of daily infections

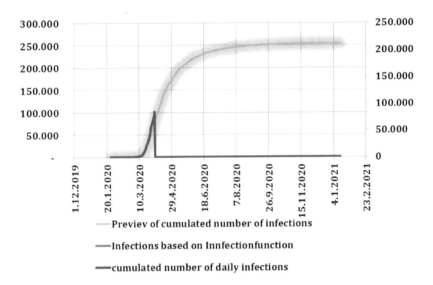

——Previev of cumulated number of infections

——Infections based on Innfectionfunction

——cumulated number of daily infections

Fig. 6.10 Infection events as of April 2nd, 2020, preview of future total development, infections based on function, total development of daily infections

The following statement is made: With the state of knowledge from the results of the measurement data on March 19, 2020, there is a forecast of 65,775 infected people by July 8, 2020.

The following is the statement: With the state of knowledge from the results of the measurement data as of March 25, 2020, there is a forecast of 123,392 infected people by July 8, 2020.

The following is the statement: With the state of knowledge from the results of the measurement data as of April 2nd, 2020, there is a forecast of 198,484 infected people by July 8th, 2020.

The following statement is made: With the state of knowledge from the results of the measurement data as of April 10, 2020, there is a forecast of 220,439 infected people by July 8, 2020.

The following statement is made: With the state of knowledge from the results of the measurement data as of May 1, 2020, there is a forecast of 208,515 infected people by July 8, 2020.

The following is the statement: With the state of knowledge from the results of the measurement data as of May 14, 2020, there is a forecast of 194,992 infected people by July 8, 2020.

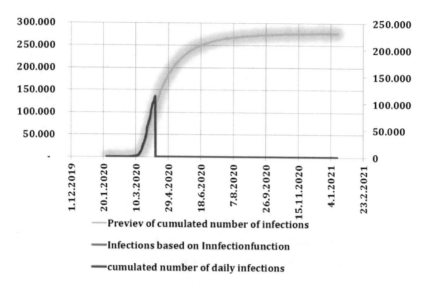

Fig. 6.11 Infection events as of April 10, 2020, preview of future total development, infections based on function, total development of daily infections

The following is the statement: With the state of knowledge from the results of the measurement data as of May 30, 2020, there is a forecast of 193,228 infected people by July 8, 2020.

From the presented diagrams, which come from the data analysis, it can be seen that the course of the infection process in the infection surveys on the intervals between the days 03/18/2020, 03/25/2020, 04/02/2020, 04/10/2020, 05/01/2020, 05/14/2020 2020, May 30th, 2020 and thus the projection into the future of the same process:

- on the date of March 18, 2020—was underestimated,
- on the date of March 25th, 2020—was increased,
- as of April 2nd, 2020—was significantly increased,
- on the date of April 10th, 2020—without a significant increase on April 2nd, 2020,
- on the date of May 1st, 2020—with a slight decrease on April 10th, 2020,
- on the date of May 1st, 2020—with a slight decrease on April 10th, 2020,
- on the date of May 14th, 2020—with a slight decrease on April 10th, 2020,
- on the date of 05/30/2020—with a slight decrease on 04/14/2020.

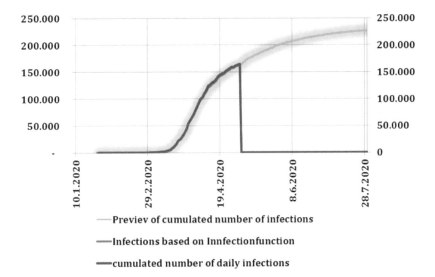

Fig. 6.12 Infection events as of May 1st, 2020, preview of future total development, infections based on function, total development of daily infections

By using the feedback from daily measurements to future events via the determination of the respective logarithm from at least 21 daily values - and thus the generation of an exponent that influences the further process and makes it predictable—we enable the basis for a probabilistic view of the future, which is made possible by the dynamics / feedback of the methods presented above.

Described specifically, all parameter values, maximum, scatter and skewness of the theoretical function change according to the dynamics / feedback described above in the course of the infection process.

6.4.3   Forecast Using the Density Function and Continuous Adjustment of the Parameters Based on a Dynamic Exponent

The used density function Equibalancedistribution (Eqb) should help to assess a forecast of the future course of infection events. As shown above, the parameters are continuously determined from the weekly mean values of the case numbers.

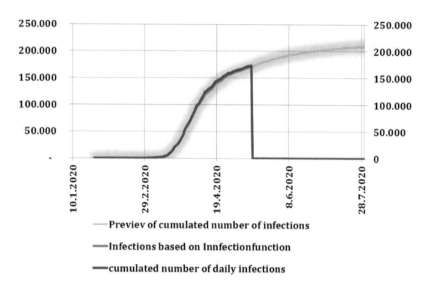

——Previev of cumulated number of infections

——Infections based on Innfectionfunction

——cumulated number of daily infections

Fig. 6.13 Infection events as of May 14, 2020, preview of future total development, infections based on function, total development of daily infections

In contrast to the determination of the prognosis for a future increase or decrease in the number of infections, a prognosis about the future frequency course and thus the expiry of the infection can be seen (Fig. 6.15 until 6.21).

he following considerations also represent graphs there that were obtained from the probability function Eqb used on the basis of the parameters from the frequency distribution.

Here, too, the statement is supported by the following diagrams for the course of the infection in Germany with the status: March 18, 2020, March 25, 2020, April 2, 2020, April 10, 2020, May 1, 2020, May 14, 2020, May 30, 2020.

For this purpose, the statistical surveys from statista.com for the states: Germany, United States of America and Spain used.

6.4.4   Germany

The following statement is made: With the state of knowledge from the results of the measurement data as of March 18, 2020, there is a forecast until the end of the infection process on 04/2020.

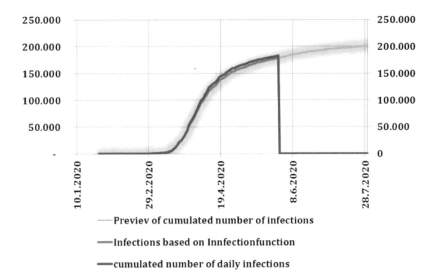

——Previev of cumulated number of infections

——Infections based on Innfectionfunction

——cumulated number of daily infections

Fig. 6.14 Infection events as of May 30, 2020, preview of future total development, infections based on function, total development of daily infections

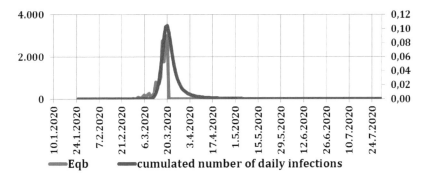

——Eqb   ——cumulated number of daily infections

Fig. 6.15 Infection events as of March 18, 2020 Frequency distribution compared to Eqb

The following is the statement: With the state of knowledge from the results of the measurement data as of March 25, 2020, there is a forecast until the end of the infection process on 05/2020.

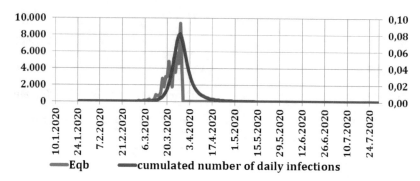

Fig. 6.16   Infection events as of March 25, 2020 Frequency distribution compared to Eqb

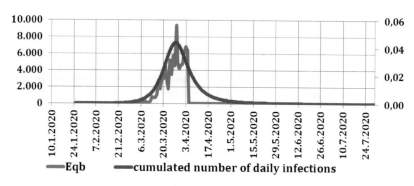

Fig. 6.17   Infection events as of 04/02/2020 Frequency distribution compared to Eqb

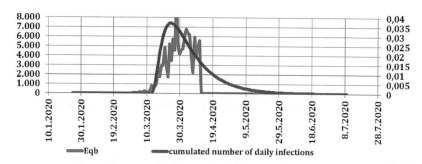

Fig. 6.18   Infection events as of 04/10/2020 Frequency distribution compared to Eqb

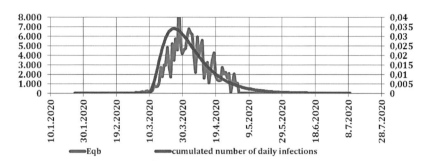

Abb. 6.19 Infektionsgeschehen Stand 01.05.2020 Häufigkeitsverteilung gegenüber Eqb

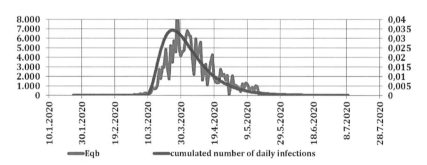

Fig. 6.20 Infection events as of May 14, 2020 Frequency distribution compared to Eqb

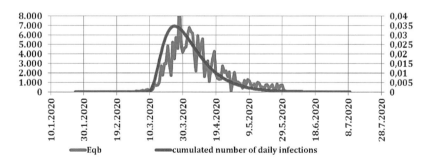

Fig. 6.21 Infection events as of May 30, 2020 Frequency distribution compared to Eqb

The following statement is made about this: With the state of knowledge from the results of the measurement data as of April 2nd, 2020, there is a forecast until the end of the infection process on 07/2020.

The following statement is made: With the state of knowledge from the results of the measurement data as of April 10, 2020, there is a forecast until the end of the infection process on 07/2020.

The following is the statement: With the state of knowledge from the results of the measurement data as of May 1, 2020, there is a forecast until the end of the infection process on 07/2020.

The following is the statement: With the state of knowledge from the results of the measurement data as of May 14, 2020, there is a forecast until the end of the infection process on 07/2020.

The following statement is made: With the state of knowledge from the results of the measurement data as of May 30, 2020, there is a forecast until the end of the infection process on 07/2020.

6.4.5 Knowledge of Germany

It is found that at the beginning of the infection process a small amount of measurement data leads to uncertainties regarding the statistical—probabilistic analyzes. The greater the number of measurements, the more obvious the end becomes.

6.4.6 United States of America

For this area the representations are made in context.

- Determination of the prognosis for a future increase or decrease in the infection rate
- Forecast using the density function and continuous adjustment of the parameters based on a dynamic exponent (Fig. 6.22 until 6.28 a,b).

The statement is supported by the following diagrams for the course of the infection in the stands: 03/18/2020, 03/25/2020, 04/02/2020, 04/10/2020, 05/01/2020, 05/14/2020, 05/30/2020.

The following statement is made: The current state of knowledge from the results of the measurement data does not provide a plausible forecast.

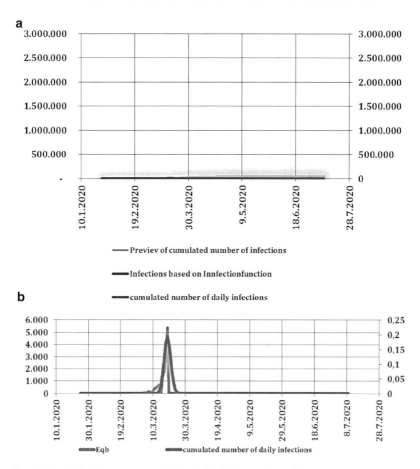

Fig. 6.22 a Infection events as of March 18, 2020, preview of future total development, infections based on function, total development of daily infections. **b** Infection events as of March 18, 2020 Frequency distribution compared to Eqb

The following statement is made about this: With the state of knowledge from the result of the measurement data, there is no forecast, since a maximum is not recognizable.

The following statement is made about this: With the state of knowledge from the result of the measurement data, there is a forecast, since a maximum is recognizable.

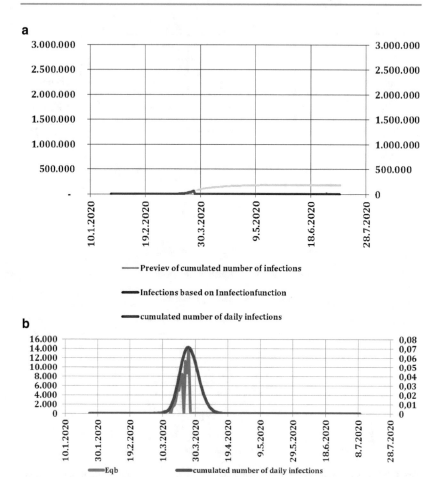

Fig. 6.23 **a** Infection events as of March 25, 2020, preview of future total development, infections based on function, total development of daily infections. **b** Infection events as of March 25, 2020 Frequency distribution compared to Eqb

The following statement is made about this: With the state of knowledge from the result of the measurement data, there is a forecast, since a maximum is recognizable.

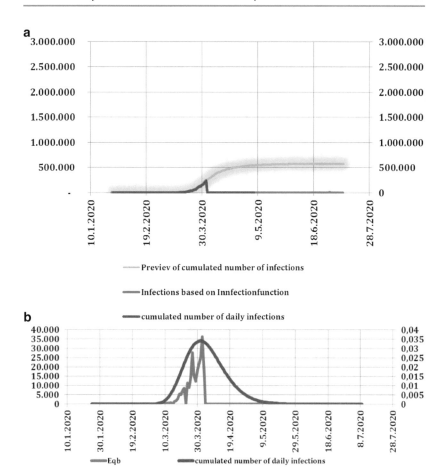

Fig. 6.24 a Infection events as of April 2nd, 2020, preview of future total development, infections based on function, total development of daily infections. **b** Infection events as of April 2nd, 2020 Frequency distribution compared to Eqb

The following is the statement: With the state of knowledge from the result of the measurement data, there is a changed forecast, since a second maximum and a significantly increased total development can be seen.

The following is the statement: With the state of knowledge from the result of the measurement data, there is a significantly changed forecast, since a second

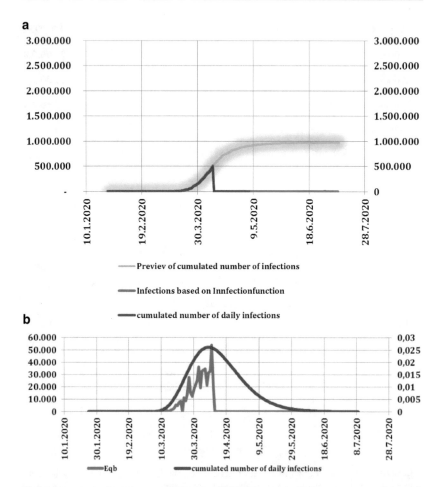

Fig. 6.25 **a** Infection events as of April 10, 2020, preview of future total development, infections based on function, total development of daily infections. **b** Infection events as of 04/10/2020 Frequency distribution compared to Eqb

maximum with a significantly variable position and a significantly increased total development can be recognized.

The following is the statement: With the state of knowledge from the result of the measurement data, there is a significantly changed forecast, since a second

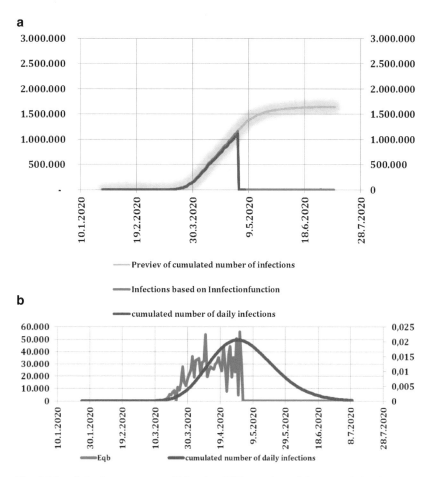

Fig. 6.26 **a** Infection events as of May 1st, 2020, preview of future total development, infections based on function, total development of daily infections. **b** Infection events as of May 1st, 2020, frequency distribution compared to Eqb

maximum with a significantly variable position and a significantly increased total development can be recognized.

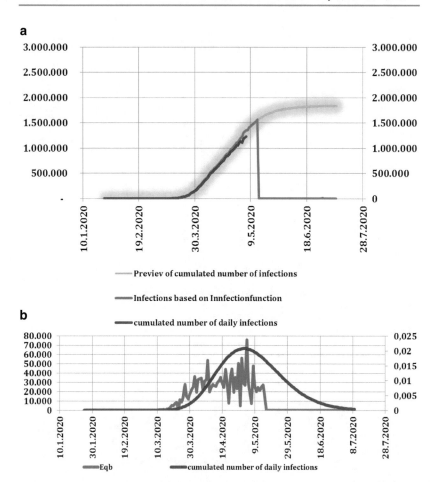

Fig. 6.27   **a** Infection events as of May 14, 2020, preview of future total development, infections based on function, total development of daily infections. **b** Infection events as of May 14, 2020, frequency distribution compared to Eqb

6.4.7   Spain

For this area the representations are made in context.

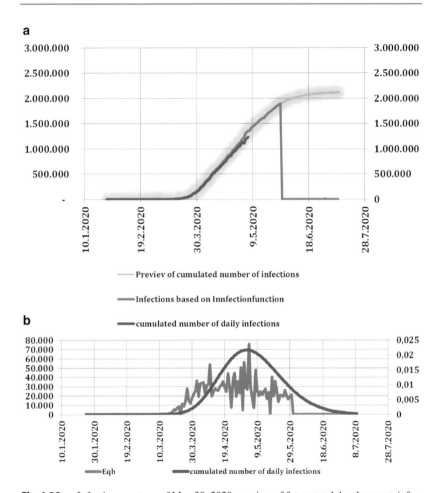

Fig. 6.28   a Infection events as of May 30, 2020, preview of future total development, infections based on function, total development of daily infections. **b** Infection events as of May 30, 2020, frequency distribution compared to Eqb

- Determination of the prognosis for a future increase or decrease in the infection rate
- Forecast using the density function and continuous adjustment of the parameters based on a dynamic exponent (Fig. 6.29 until 6.35 a,b)

a

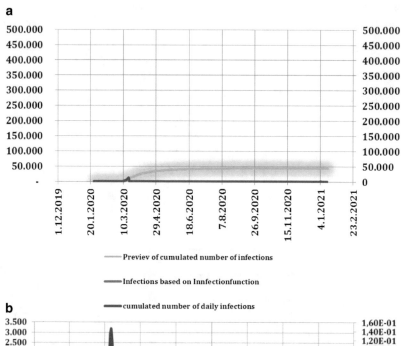

───── Previev of cumulated number of infections

───── Infections based on Innfectionfunction

───── cumulated number of daily infections

b

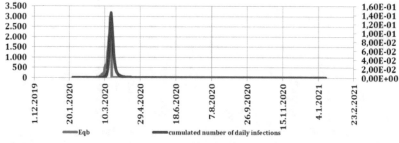

Fig. 6.29 **a** Infection events as of March 18, 2020, preview of future total development, infections based on function, total development of daily infections. **b** Infection events as of 03/18/2020 Frequency distribution compared to Eqb

The statement is supported by the following diagrams for the course of the infection in Germany with the statuses: March 18, 2020, March 25, 2020, April 2, 2020, April 10, 2020, May 1, 2020, May 14, 2020, May 30, 2020.

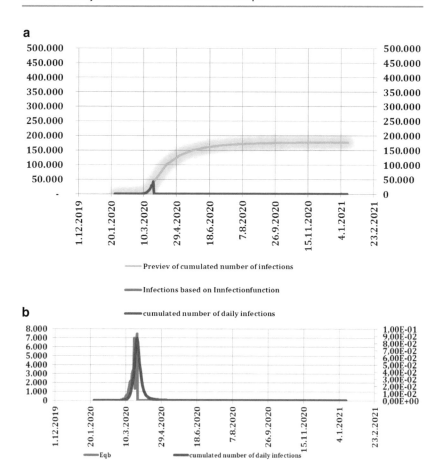

Fig. 6.30 **a** Infection events as of March 25, 2020, preview of future total development, infections based on function, total development of daily infections. **b** Infection events as of March 25, 2020 Frequency distribution compared to Eqb

The following statement is made: With the state of knowledge from the results of the measurement data, there is an extremely steep increase in the number of cases without a maximum being apparent.

The following statement is made: With the state of knowledge from the results of the measurement data, there is an extremely steep increase in the—increasing—number of cases without a maximum being apparent.

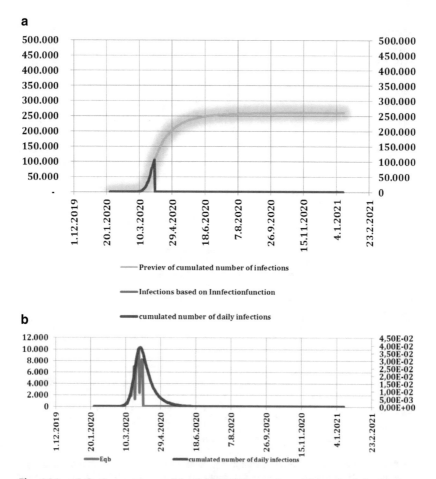

Fig. 6.31 **a** Infection events as of April 2nd, 2020, preview of future total development, infections based on function, total development of daily infections. **b** Infection events as of April 2nd, 2020 Frequency distribution compared to Eqb

The following statement is made about this: With the state of knowledge from the results of the measurement data, there is an extremely steep increase in the—still increasing—number of cases without a maximum being apparent.

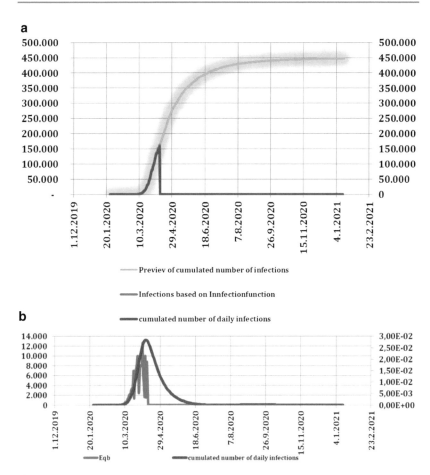

Fig. 6.32 **a** Infection events as of April 10, 2020, preview of future total development, infections based on function, total development of daily infections. **b** Infection events as of April 10, 2020 Frequency distribution compared to Eqb

The following statement is made about this: With the state of knowledge from the results of the measurement data, there is an extremely steep increase in the— still increasing—number of cases, the second turning point becomes apparent.

The following is the statement: With the state of knowledge from the result of the measurement data, there is a decrease in the number of cases, the second turning point becomes apparent.

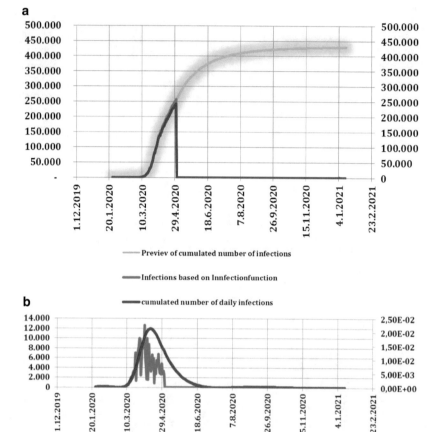

Fig. 6.33 a Infection events as of May 1st, 2020, preview of future total development, infections based on function, total development of daily infections. **b** Infection events as of May 1st, 2020 frequency distribution compared to Eqb

The following statement is made: With the state of knowledge from the results of the measurement data, the number of cases will decrease and an end is foreseeable.

The following statement is made: With the state of knowledge from the results of the measurement data, the number of cases will decrease and an end is foreseeable.

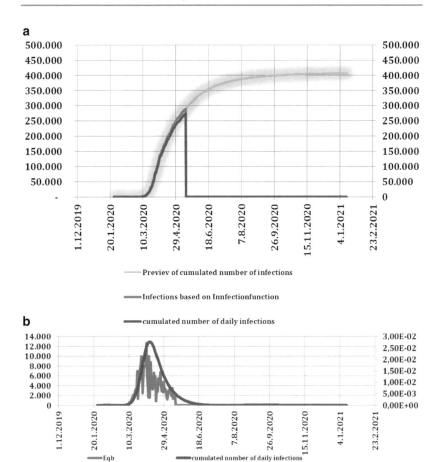

Fig. 6.34 **a** Infection events as of May 14, 2020, preview of future total development, infections based on function, total development of daily infections. **b** Infection events as of May 14, 2020 frequency distribution compared to Eqb

6.5 Consideration of Some Developments in the United States

In the previous considerations, the measurement data from countries were considered. As such, they must be viewed as a swarm of data because the data does not come from a single population. The data collections from https://github.com/nyt

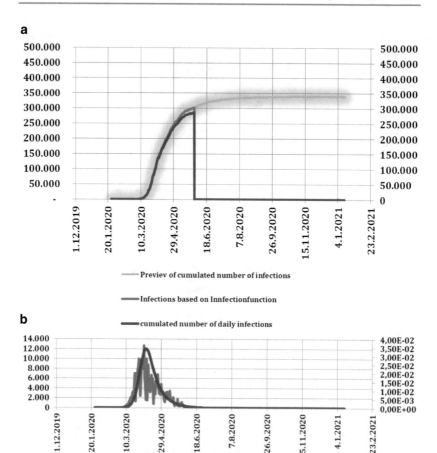

Fig. 6.35 **a** Infection events as of May 30, 2020, preview of future total development, infections based on function, total development of daily infections. **b** Infection events as of May 30, 2020 Frequency distribution compared to Eqb

imes/covid-19-data/blob/master/us-states.csv allow the results from measurement data for some of the US states to be viewed.

Based on the considerations of some developments in the USA it will now be shown (Fig. 6.36, a, b, c, d, e until 6.48):

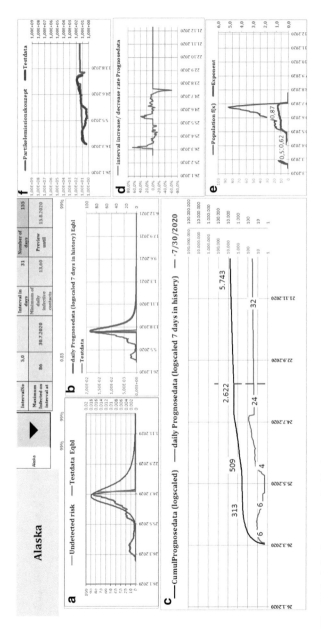

Fig. 6.36 Infection process in Alaska

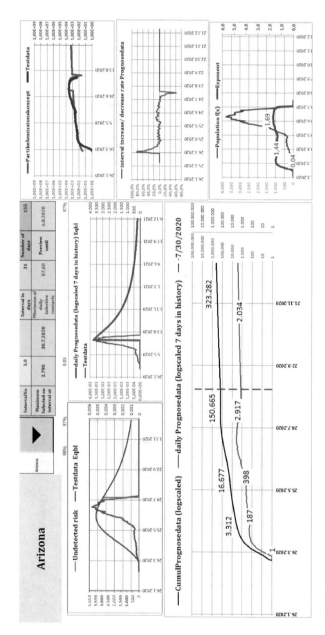

Fig. 6.37 Arizona infection

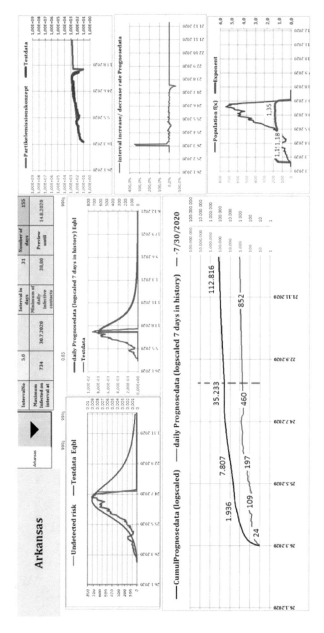

Fig. 6.38 Infections in Arkansas

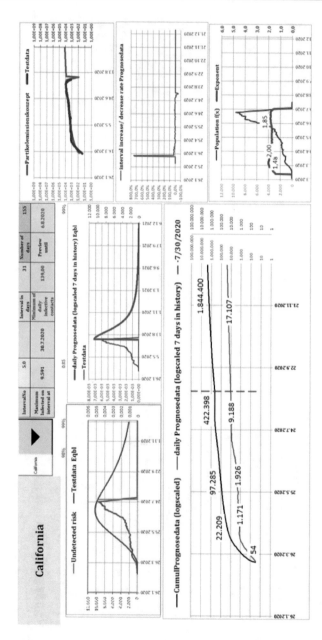

Fig. 6.39   Infection events in California

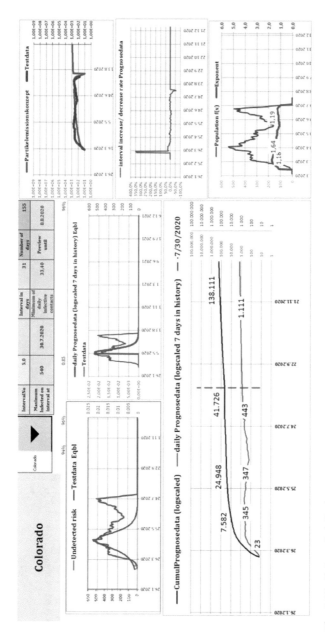

Fig. 6.40 Col or ado infection

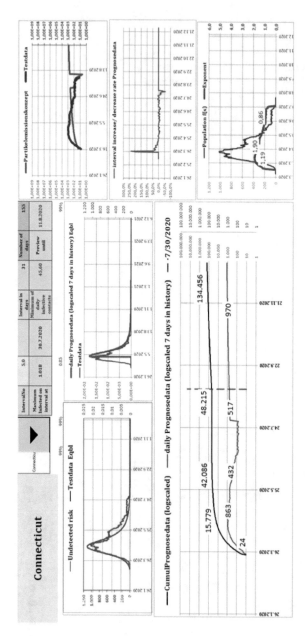

Fig. 6.41   Connecticut infection

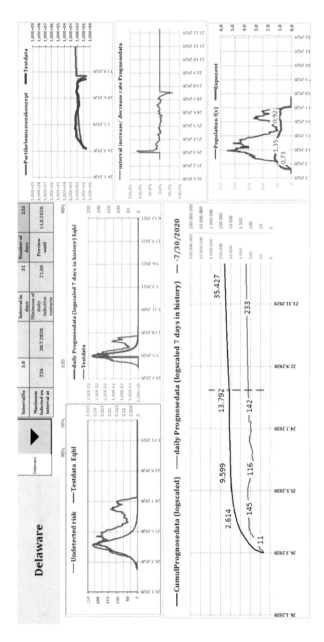

Fig. 6.42 Delaware infection

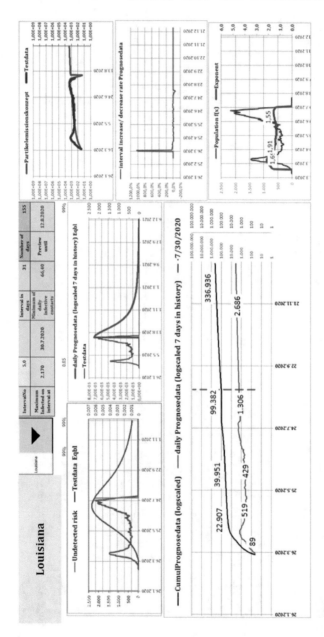

Fig. 6.43   Luisiana infection process

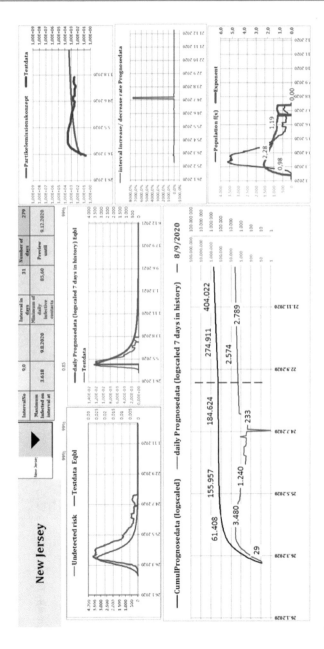

Fig. 6.44 Infection in New Jersey

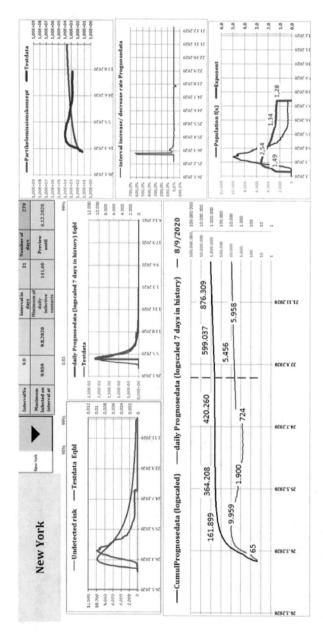

Fig. 6.45   Infection in New York

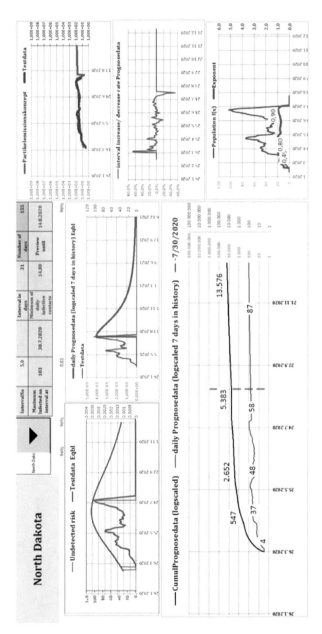

Fig. 6.46 Infections in North Dakota

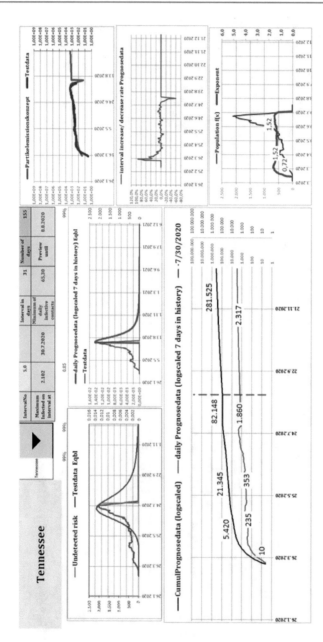

Fig. 6.46 Tennessee infection

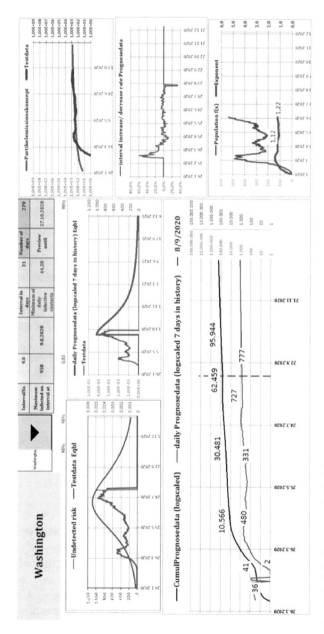

Fig. 6.47 Infection events in Washington

1. The presentation of the expected future development and thus the undiscovered risk according to the probability function,
2. The conclusion from the logarithmic development from a 1-week period from the anticipatory development on the future exponent according to the probability function,
3. The prognosis of the course of the infection daily individually and in total,
4. The increase or decrease in infection as a percentage,
5. Determining the exponent of the current development,
6. The comparison of the development of the frequency from the test data with the values from the particle emission concept

Based on actual data from COVID-19, it shows: Some graphs of the development in some US states of good behavior and bad behavior over time. Most important to see: The exponent is based on the mean logarithm of the 7 days before it increased / decreased.

The following statement is made about this: With the state of knowledge from the result of the measurement data, **the number of cases increases if the exponent continues to rise.**

The following statement is made about this: With the state of knowledge from the result of the measurement data, **the case numbers are equal if the exponent remains unchanged.**

The following statement is made about this: With the state of knowledge from the result of the measurement data, **the case numbers are equal if the exponent remains unchanged.**

The following statement is made about this: With the state of knowledge from the result of the measurement data, **the case numbers are equal if the exponent remains unchanged.**

The following statement is made about this: With the state of knowledge from the result of the measurement data, the number of cases increases if the exponent continues to rise. **A second increase shows the inattention to the unknown propagation properties.**

The following statement is made about this: With the current state of knowledge from the results of the measurement data, **the number of cases decreases because the exponent decreases continuously and remains stable.**

The following statement is made about this: With the current state of knowledge from the results of the measurement data, the number of cases decreases because the exponent decreases continuously and remains stable. **A second increase shows the inattention to the unknown propagation properties.**

The following is the statement: With the state of knowledge from the results of the measurement data, there is an **increase in the number of cases because the exponent increases continuously. A second increase shows the inattention to the unknown propagation properties.**

The following statement is made about this: With the current state of knowledge from the results of the measurement data, **the number of cases decreases because the exponent decreases continuously and remains stable.**

The following statement is made about this: With the current state of knowledge from the results of the measurement data, **the number of cases decreases because the exponent decreases continuously and remains stable.**

The following is the statement: With the state of knowledge from the results of the measurement data, there is an increase in the number of cases because the exponent increases continuously. **A second increase shows the inattention to the unknown propagation properties.**

The following statement is made about this: With the state of knowledge from the result of the measurement data, **the case numbers are equal if the exponent remains unchanged.**

The following is the statement: With the state of knowledge from the results of the measurement data, there is an increase in the number of cases because the exponent increases continuously. **A second increase shows the inattention to the unknown propagation properties.**

6.6 Incidence Under a Probabilistic View

In most cases, the "incidence" means the "cumulative incidence". This is a measure of how many healthy people of a defined population group (e.g. all residents of Berlin / Germany / Europe; all 18 to 25-year-old male, non-smoking Australians) fall ill with a certain disease within a certain period of time. People who were already sick or who were sick again before this period are not included in the calculation.

$$I = \frac{\text{Number of new cases in a population over a given period of time}}{\text{sum of all times of all individuals to become ill}} \quad (6.11)$$

Example given:

121 new cases were reported over 10 years based on a population of 111.000 healthy people.

How is the (cumulative) incidence per year related to 100,000 people?

- 121/111.000 ->0,00,109,009 for 10 years
- Per year: 0,00,109,009
- Per 100,000 people: 109
- I.e. 121 new cases per 100,000 develop each year.

The cumulative incidence can be used to estimate the likelihood that a person from the group of people under consideration will develop the disease.

The upper work may demonstrate, that the evolution of an incidence may be completed by a probabilistic view, when the number of cases increases exponentially. Also the lonely exponential view may be completed by a view via a density function that offers a more detailed contemplation by respecting the number of cases some days in history that give a logarithm value as an exponential forecast—base.

Based on the development of the data, which correspond to the probability calculation, the following expected value for the incidence of one week can be assumed. For this purpose, all values of the density Eqbl are determined from the parameter values modal value, standard deviation, skewness and kurtosis from the previous week for the future week according to the formula:

$$Eqb4(x; \delta, md, r, \kappa) = (\frac{1}{s * \sqrt{\left(2\pi\left(\frac{1-((r)*(x-md))}{\kappa}\right)\right)}} * EXP\left(\left(-\left(\frac{1}{2} * \frac{\left(\frac{x-md}{s}\right)^2}{1-(r*(x-md))}\right)*\kappa\right)\right)$$

(6.6.2)

The effects to be expected if the statement from

Zhongwei Cao1, Xu Liu1, Xiangdan Wen2, Liya Liu3 & Li Zu4,
A regime-switching SIR epidemic model with a ratio-dependent incidence rate and degenerate diffusion
In the SIR model, individuals are in one of three states: S = susceptible, I = infected, R = removed (cannot be infected).

Note: In the present work, only test data from demonstrably infected groups (I) were considered, a probabilistic analysis for groups S and R must therefore be carried out separately and with the corresponding populations!
Quote:

"Based on the pioneering research1, mathematical model provides effective control measures for infectious diseases and is an significant tool for analyzing the epidemiological characteristics of infectious diseases. In the course of the spread of disease, the transmission function plays an important role in determining disease dynamics.

There are several nonlinear transmission functions proposed by authors. For instance, Capasso and Serio introduced a saturated incidence rate g(I)S into epidemic models, and the infectious force $g(I)$ is a function of an infected individual that is applied in many classical disease models. Liu et al. proposed a general incidence rate:

$$g(I)S = \beta(I)^{P}S/(1 + pI^{q}), \, p, q > 0$$

Lahrouz et al. introduced a more generalized incidence rate

$$g(I)S = SI/f(I).$$

In particular, Yuan and Li8 considered a ratio-dependent nonlinear incidence rate in the following
Form:

$$g(I/S)S = \beta(I/S)^{l}S/(1 + \alpha(IS)^{h}))$$

where α is a parameter used to measure psychological or inhibitory effects.
Since statistics and probability calculations can possibly have a positive influence on the situation, the suggestion is made that the parameter α should no longer measure psychological or inhibitory effects, but rather serve as the expected probability value, so that at every previous weekly interval (7 days for COVID 19) a probabilistic statement is available as a preview.

6.6.1   Probabilistic Incidence preview for Texas

A probabilistic preview of the incidence (Fig. 6.49) is possible at short notice, especially if the sum of the 7 daily values is constant. Since the course of the infection process always runs exponentially, the previous logarithm of the infection process must be observed, because this has a decisive influence on the exponent for the forecast. To demonstrate this connection, we use a time segment of the US state of Texas between May 25th. and June 3rd. A probabilistic incidence of approx. 30.9% can then be expected. The incidence calculation is based on the following:

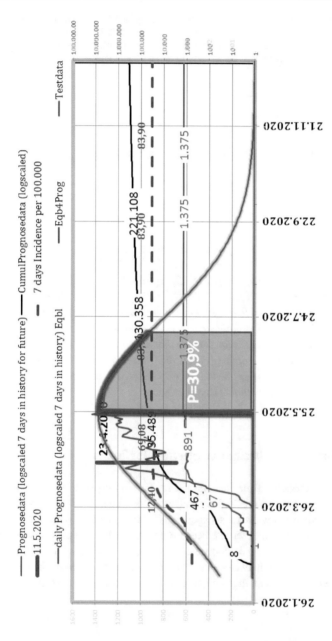

Fig. 6.49 Probability of the Incidence in between 2 daily dates

$$I = \frac{\text{number of new cases in a population within 7 days}}{\text{Number of residents of a state}} * 10 \,^{\wedge} 5 \qquad (6.6.3)$$

Leakage Effect—Percolation of the Virus 7

When groups of living beings—people are one of them—come together, particles are exchanged in accordance with the particle emission concept. The particles are transmitted or exchanged via those infected who have:

- the skin, mucous membrane (handshake, greeting kiss, exchange of things, such as glasses, bottles and the like)
- breathing (singing, speaking, heavy breathing)
- are in close spatial relationship with one another

so that an exchange of particles—the closer the distances are to each other

- becomes more frequent and more intense!

7.1 Potential Implications for Health Care Settings and Epidemiological Modeling

7.1.1 Beta Coefficient in Nonlinear Epidemiological Modeling

One of the most important potential implications of the current probabilistic model proposed here is that it could serve as a basis for correct estimates of Beta used in nonlinear dynamic modeling. For example, in SIR models the normal distribution for the estimates of error could be replaced with the Eqb distribution. Estimating Beta is key as most of the other model dynamics are influenced by this infection per interaction probability, and so getting the most correct estimation is

of the ultimate benefit of the entire model's accuracy and efficacy. More reali-
stic parameter estimates could lead to better modeling which would not need to
be exceedingly complex but which would be more useful given a more accurate
disease transmission coefficient. With a less accurate coefficient, even factoring
in hundreds of variables may lead to less accurate results than a model with fewer
variables but a more accurate Beta.

On this subject is listed:

Frank Ball, Tom Britton, Ka Yin Leung & David Sirl, "A stochastic SIR
network epidemic model with preventive dropping of edges".

"10.2 Convergence and approximation of temporal properties First we demonstrate
numerically some of the limit theorems from earlier sections, showing both how the
convergence is realised and thus how these limit theorems can be used for approximation. We give examples only with an NSW graph construction, but much the same
observations apply in the MR graph scenario.

In Fig. 1 we demonstrate using Theorem 7.2 for approximation of the temporal evolution of the epidemic, comparing simulated trajectories of the prevalence I N (t) (for N
= 1000) versus time t of the model with predictions from the functional central limit
theorem, for a Poisson and a Geometric degree distribution. The upper plots show
the simulated trajectories together with the mean and a central 95% probability band
predicted by the CLT; they suggest that the approximation is fairly good. The lower
plots compare the mean and standard deviation of the prevalence through time with
the LLN and CLT based asymptotic predictions.

In Fig. 2 we investigate the convergence of the distribution of I N (t) to its N → ∞ limit at
three time points t1, t2 and t3. The times are chosen so that t2 is close to the time of peak
prevalence and t1 and t3 are when prevalence is increasing and decreasing,respectively,
at a level roughly half that of the peak prevalence. (Effectively we are examining
the upper-right plot of Fig. 1 in detail at these three time points.) In this figurewe
have used a geometric degree distribution, but very similar conclusions are obtained
using different distributions. This convergence is further investigated/demonstrated
in Figure 7.1, where, separately for each of the same three time points, we plot the
Kolmogorov distance between the empirical and asymptotic distributions of the number
of infectives against population size N."

In the previous presentation, reference is made to the demonstration of the
approximation of the measured values from surveys in frequency distributions
to Poisson and geometric distribution. The Equibalancedistribution can certainly
be an alternative for such approximations, especially since a regression analysis of the least squares between the frequency distribution and the Equibalance
distribution achieves very high measures of determination (Fig. 7.2).

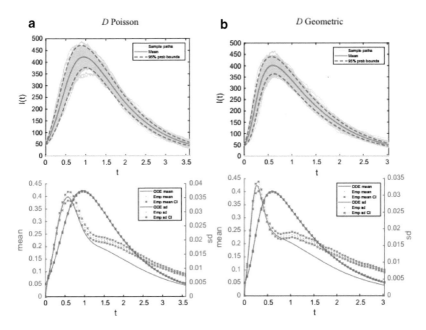

Fig. 7.1 Demonstration of approximation implied by theorem 7.2

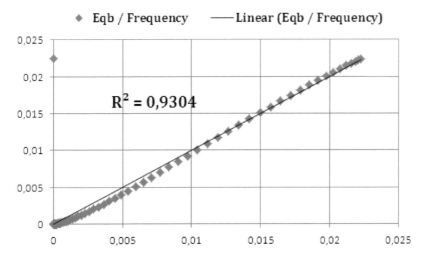

Fig. 7.2 Measures of determination Eqb/frequency

7.1.2   Health Care Setting Management

In addition to the potential for modeling, the proposed method and the percolation effect could be considered in health care settings where frequent and repeated contacts could result in higher viral loads per individual which could then spread throughout the health care environment.

Given this consideration, rotations of personnel (when possible) could help to mitigate loads and contact points.

A concrete example would be in an emergency department where there is typically a high volume of patients and health care workers, and where the ability to control respiratory droplets and aerosols may be challenging especially given the close quarters of the facilities.

If there were enough personnel, it is conceivable that they personnel could be rotated to work in different areas with different patients in shifts that would limit exposure and decrease viral loads. This may be impractical in many situations, but the concept is worth exploring where possible.

Perhaps another approach would be the use of open-air triage hospitals such as the field hospitals that were set up in the United States in many locations. Increasing the space could allow for decreased viral loads assuming that the air ventilation and increased distance per patient (unless contact is absolutely necessary) and may reduce the percolation effect.

The use of the probability calculation in connection with the early detection of the increase in the infection rate by determining the exponent from the logarithm of the previous test data should help:

- to dampen the evolution at an early stage through suitable measures,
- to make the SIR modeling more precise by replacing an expected symmetrical distribution around a mean value by an asymmetrical one that indicates extreme developments
- and thereby the percolation is largely prevented.

7.2   On the Percolation Theory COVID

7.2.1   A Basic Consideration, Mold Percolation

An act from the natural environment of a household should give a first insight into the percolation—the leakage—of infections in a population:

There is a collection of freshly collected, moist walnuts in a basket. They are spread out on a table to be counted (Fig. 7.3).

At the time of counting, they are close together and will go moldy if not separated (Fig. 7.4).

The walnuts are all separated to prevent mold formation (Fig. 7.5).

The basis for a consideration of the distribution lies in the group formation of elements, in this case walnuts, shown here in 4 groups (Fig. 7.6), each of which comes into contact with an element of another group and is also in contact with everyone within the group. One group is not in contact with another. The mold infection spreads within the group and is transmitted to another group.

Fig. 7.3 A collection of walnuts

Fig. 7.4 A collection of walnuts are close together

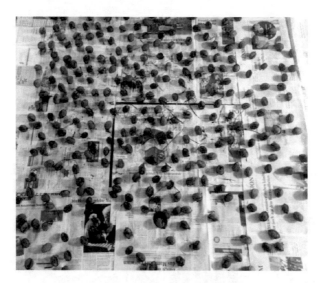

Fig. 7.5 A collection of walnuts lie at a distance from each other

Fig. 7.6 Groups in contact, not in contact

If groups are gathered in the immediate vicinity, mold growth can spread over the entire population, as each element of one group is in contact with many elements in other groups (Fig. 7.7).

7.2.2   Consideration of the Vius Percolation in Human Populations

What does mold growth on walnuts have in common—in principle—with an infection, what is the difference between them?

– A walnut has a single surface, the shell that surrounds and protects the nut fruit,
– A living being like humans (or any other living being) has a surface that runs continuously from the outside to the inside
– the inner surface—and is supposed to protect the organism on
– the inside via mucous membranes.

Fig. 7.7 Many groups in contact

In this respect, humans are also exposed to loads from the air they breathe due to their large surface and inner surface, which can contain pollutants—viruses.

Theories of percolation have been presented in various theoretical treatises. The mathematical expression is described as follows: https://de.wikipedia.org/wiki/Perkolation:

• "Percolation theory, a mathematical model of cluster formation in grids"

or the application "percolation of the COVID -19 system", the infection process is subject to conditions and parameters in a model calculation which are named as follows:

7.2.3   Conditions for a COVID Model Calculation

7.2.3.1 Initial Cluster/Cluster—Grid

Conditions: A number of initial clusters form independently of one another in a nucleus which can have up to n elements (infected). A cluster is defined by the number of edges, which are given by the definition in the wording:

"Every 1 element (participant in an initial group) shares an infection with all other participants at an initial point in time without contact to another group".

A number of initial clusters form a successor cluster grid.

7.2.3.2 Follower Cluster/Follower Cluster Grid

Conditions: A number of sequential clusters form an interdependent grid that can have up to n contacting elements (infected). A sequence cluster grid is defined by the number of edges, which are given by the definition in the wording:

"At least 2 elements (participants from subsequent clusters) share an infection with all other participants at a subsequent point in time with contact to another group".

Examples of the initial clusters are shown as follows (Fig. 7.8):

The following formula applies to the groups for the number of the minimum number of particle transfers as stated above.

$$- A_{E,2} = (n_E * (n_E - 1))/27.2$$

This results in the following transmission rate for the groups:

a) $A_{E,3} = (3 * 3 - 1)/2 = 3, = {>}3$ individuals contact at least 3 times $= 3/3 = 100\%$

b) $A_{E,4} = (4 * 4 - 1)/2 = 6, = {>}4$ individuals contact at least 6 times $= 6/4 = 150\%$

c) $A_{E,5} = (5 * 5 - 1)/2 = 10, ={>}5$ individuals contact at least 10 times $= 10/5 = 200\%$

d) $A_{E,6} = (6 * 6 - 1)/2 = 15, ={>}6$ individuals contact at least 15 times $= 15/6 = 250\%$

The percolation effect sets in when the following occurs:

a   b   c   d

Fig. 7.8 Independent groups of a 3, b 4, c 5, d 6 individuals, Examples of Initial—Clusters

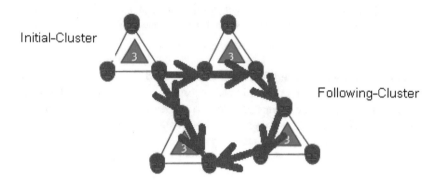

Fig. 7.9   A beginning percolation cluster

- The first percolation effect occurs in an initial group
- Each follow-up group from an Initial-group contacts each other repeatedly.
- Following-Clusters form grids (Fig. 7.9)

7.2.3.3 Beginning of the Percolation by Breathing (Singing, Speaking, Heavy Breathing

This leads to the following observation:

- many individuals receive a multiple of the original transmission rate.
- many individuals contacted many individuals
- during the observation period, the viruses multiply according to the organs provided

As soon as the percolation limit is reached, the transfer structure changes into a

- chaotic system, which is no longer recognizable in its structure.

The percolation limit is reached when at least—initially independent groups—come together.

    The development of the system can no longer be traced and is therefore chaotic (Fig. 7.10).

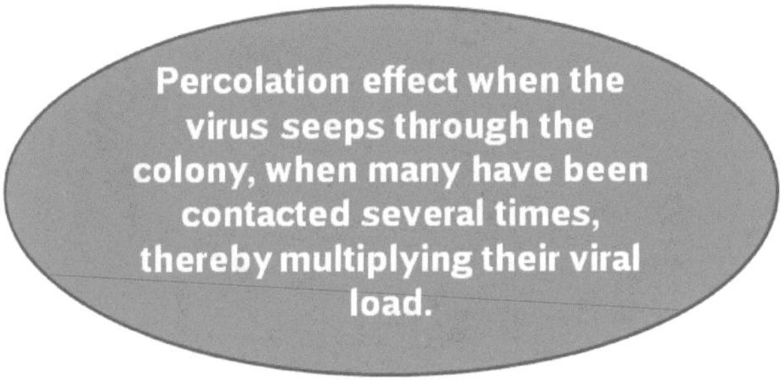

Percolation effect when the virus seeps through the colony, when many have been contacted several times, thereby multiplying their viral load.

Fig. 7.10   Percolation effect

7.3   Principles of Percolation- Interface Effects

The principle of percolation of infections of groups develops from the superposition of the grids according to the above pattern (Fig. 7.11). The contact points from the grids add up completely over the common interfaces until after one

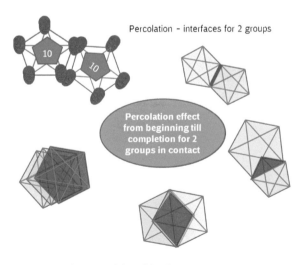

Percolation – interfaces for 2 groups

Percolation effect from beginning till completion for 2 groups in contact

Leakage effect - percolation of the virus

Fig. 7.11   Percolation interfaces

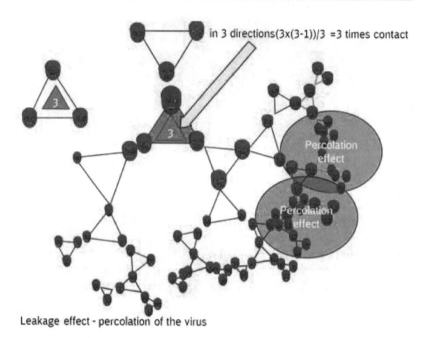

Leakage effect - percolation of the virus

Fig. 7.12   Percolation effect, in 3 directions $(3 \times (3 - 1))/2 = 3$ times contact

period—in the case of infections. In the case of a large number of groups that come into contact with each other (Fig. 7.12, 7.13, 7.14 and 7.15), the traceability of the infection process is no longer traceable, since the individual is infected several times.

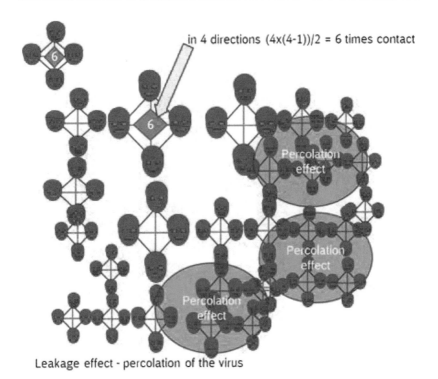

Fig. 7.13   Percolation effect, in 4 directions $(4 \times (4 - 1))/2 = 6$ times contact

7.4   Examples of Percolation Effects, Clustering

7.4.1   Table of Initial-Cluster Without Percolation Effect

(See Fig. 7.16)

7.4.2   Initial-Cluster/Follwing-Cluster with a Percolation Effect

As shown in the table under 7.16, an exponential calculation is not sufficient to determine the expected number of cases—it appears rather in a hyper-exponential way, depending on

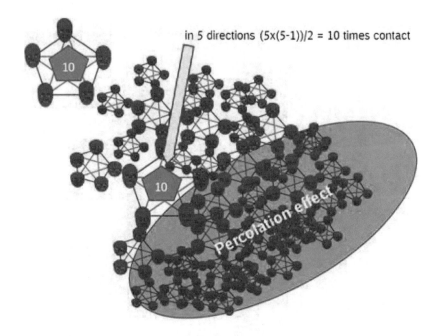

Leakage effect - percolation of the virus

Fig. 7.14 Leakage effect, in 5 directions $(5 \times (5 - 1))/2 = 10$ times contact

- the way in which the initial clusters and subsequent clusters are developed
- the number of people they are occupied
- the number of all clusters working together—in a grid.

As a result, a computational forecast is only possible if a grid formation is avoided, otherwise a forecast can only be made in a probabilistic manner.

7.5   The Consequences of the Percolation Effect, Germany

The consequences of the percolation effect develop chaotically in a system that is no longer comprehensible.

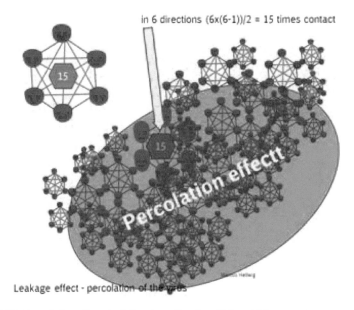

Fig. 7.15 Percolation effect, in 6 directions $(6 \times (6 - 1))/2 = 15$ times contact

Up until May 2020, the exponent of a development using the logarithm of the previous 7 days was transferrable to the frequency distribution of the exponential infection process in Germany (Fig. 7.17, 7.18, 7.19 and 7.20).

Compared to other countries, Germany had—obviously—a controlled, decreasing course of the infection process.

Even a spontaneous increase over a short period of time did not initially have a great influence on the overall course, but the observations of the logarithmic 7-day development had indicated that the percolation has a beginning.

By October at the latest, it became apparent that the infection process could no longer be traced, the percolation prevented traceability.

It is therefore foreseeable that the infection process will not end before the middle of 2021 (Fig. 7.21).

The datasets from the test data of the federal states (Fig. 7.17, 7.18, 7.19 and 7.20).provide information about which population has most consistently adhered to the hygiene rules and what time interval can be expected before the infection process comes to an end (Fig. 7.22 and 7.23).

Number of persons		Number of the day 1 contacts when a number of people meet	Number of the day 2 contacts when a number of people meet	Number of the day 3 contacts when a number of people meet	Number of the day 4 contacts when a number of people meet	Total cases in 4 days
0		0	0	0	0	0
1		0	0	0	0	0
2		1	0	0	0	1
3		3	3	3	3	12
4		6	15	105	5.460	5.586
5		10	45	990	489.555	490.600
6		15	105	5.460	14.903.070	14.908.650
7		21	210	21.945	240.780.540	240.802.716
8		28	378	71.253	2.538.459.378	2.538.531.037
9		36	630	198.135	19.628.640.045	19.628.838.846
10		45	990	489.555	119.831.804.235	119.832.294.825
11		55	1.485	1.101.870	607.058.197.515	607.059.300.925
12		66	2.145	2.299.440	2.643.711.007.080	2.643.713.308.731
13		78	3.003	4.507.503	10.158.789.393.753	10.158.793.904.337
14		91	4.095	8.382.465	35.132.855.546.880	35.132.863.933.531
15		105	5.460	14.903.070	111.050.740.260.915	111.050.755.169.550
16		120	7.140	25.486.230	324.773.947.063.335	324.773.972.556.825
17		136	9.180	42.131.610	887.536.259.530.245	887.536.301.671.171

Fig. 7.16 Table of leakage effects up to 17 persons

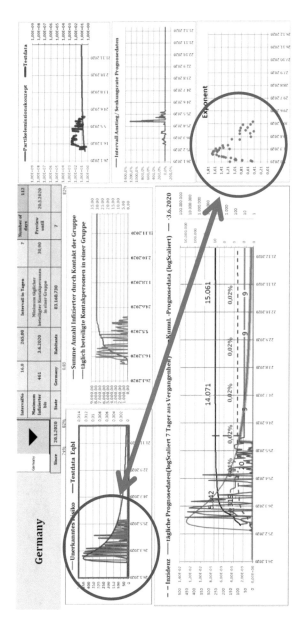

Fig. 7.17 Germany May-2020, Testing data, daily infection rate, Exponent of exponential infection

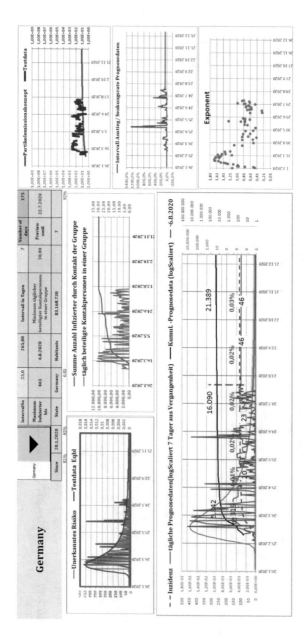

Fig. 7.18 Germany August-2020, Testing data, daily infection rate, Exponent of exponential infection

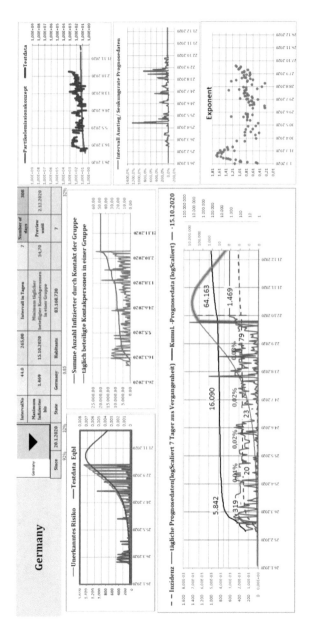

Fig. 7.19 Germany October-2020, Testing data, daily infection rate, Exponent of exponential infection

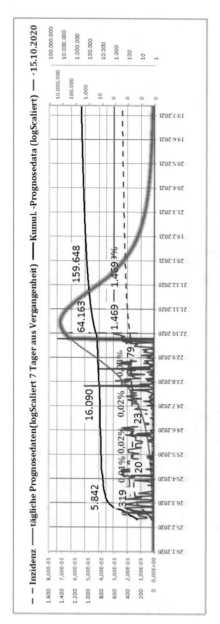

Fig. 7.20 Germany October-2020, Testing data, daily infection rate, Exponent of exponential infection

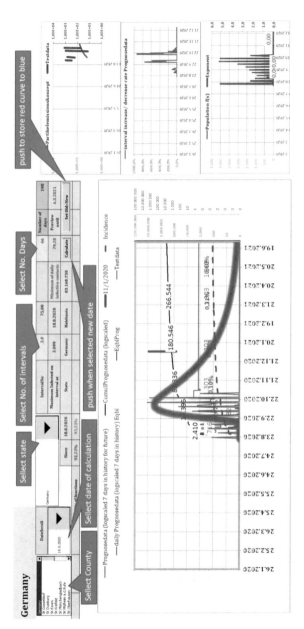

Fig. 7.21 Germany November-2020, Testing data, daily infection rate, Exponent of exponential infection from the system of the author

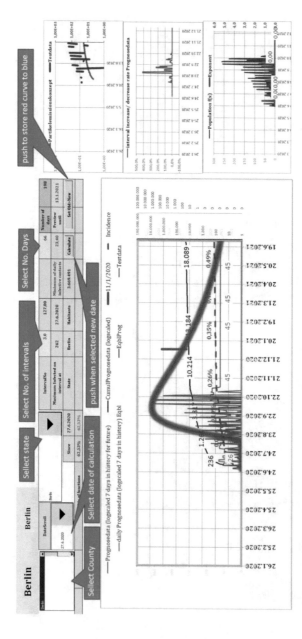

Fig. 7.22 Berlin November-2020, Testing data, daily infection rate, Exponent of exponential infection from the system of the author

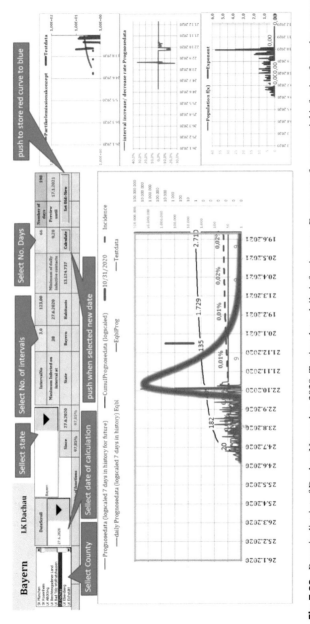

Fig. 7.23 Bavaria district of Dachau November-2020, Testing data, daily infection rate, Exponent of exponential infection from the system of the author

7.6   Summary

The preceding chapters may explain that infection processes can be analyzed systematically using the known methods from statistics, stochastics and probability theory. The case of biological infection was considered in the present study. The same methods can be applied to other areas; this means any infection process, including the digital one. The matter to be considered equally in all areas is the earliest possible detection of an "attack" on the organism to be protected, be it biological, digital or organizational structure, of any kind can increase rapidly in their frequency at the beginning if they cannot be recognized early and moderate or even prevented. Infection risks remain active as long as they are described, observed and measured, as long as they are not recognized and without resistance. The graphs show the multi- exponential multiplication in which these develop and flare up again if the emission paths are not consistently interrupted.

The modeling of an infection process using the S.I.R. methods described can benefit from the fact that the parameters for skewness and kurtosis taken into account in the equi-balance distribution take into account the actual skewness and steepness of an exponential expression of the frequency distribution, a good approximation of the asymmetrical course over time.

7.7   Bibliography/Source Information

– Federal Government Gemany and Robert Koch Institute, data base: www.sta tista.com,
– Zhongwei Cao1, Xu Liu1, Xiangdan Wen2, Liya Liu3 & Li Zu4ations of the Federal,
  "A regime-switching SIR epidemic model with a ratio-dependent incidence rate and degenerate diffusion"
– https://www.rki.de/DE/Content/InfAZ/N/Neuartiges_Coronavirus/Situationsbe richte/2020-07-29-de.pdf?__blob=publicationFile
   – Frank Ball, Tom Britton, Ka Yin Leung & David Sirl, Journal of Mathematical Biology volume 78, pages 1875–1951(2019); A stochastic SIR network epidemic model with preventive dropping of edges

Printed in the United States
by Baker & Taylor Publisher Services